Moral Hermeneutics and Technology

Postphenomenology and the Philosophy of Technology

Editor-in-Chief: Robert Rosenberger

Executive Editors: Peter-Paul Verbeek and Don Ihde

As technologies continue to advance, they correspondingly continue to make fundamental changes to our lives. Technological changes have effects on everything from our understandings of ethics, politics, and communication, to gender, science, and selfhood. Philosophical reflection on technology can help draw out and analyze the nature of these changes, and help us to understand both the broad patterns of technological effects and the concrete details. The purpose of this series is to provide a publication outlet for the field of philosophy of technology in general, and the school of thought called "postphenomenology" in particular. The field of philosophy of technology applies insights from the history of philosophy to current issues in technology and reflects on how technological developments change our understanding of philosophical issues. Postphenomenology is the name of an emerging research perspective used by a growing international and interdisciplinary group of scholars. This perspective utilizes insights from the philosophical tradition of phenomenology to analyze human relationships with technologies, and it also integrates philosophical commitments of the American pragmatist tradition of thought.

Recent Titles in This Series

Moral Hermeneutics and Technology: Making Moral Sense through Human-Technology-World Relations by Olya Kudina

Postphenomenology and Imaging: How to Read Technology, edited by Samantha J. Fried and Robert Rosenberger

Postphenomenology and Architecture: Human Technology Relations in the Built Environment, edited by Lars Botin and Inger Berling Hyams

Technology and Anarchy: A Reading of Our Era, by Simona Chiodo

How Scientific Instruments Speak: Postphenomenology and Technological Mediations in Neuroscientific Practice by Bas de Boer

Feedback Loops: Pragmatism about Science & Technology, edited by Andrew Wells Garnar and Ashley Shew

Sustainability in the Anthropocene: Philosophical Essays on Renewable Technologies, edited by Róisín Lally

Unframing Martin Heidegger's Understanding of Technology: On the Essential Connection between Technology, Art, and History by Søren Riis, translated by Rebecca Walsh

Moral Hermeneutics and Technology

Making Moral Sense through Human-Technology-World Relations

Olya Kudina

LEXINGTON BOOKS
Lanham • Boulder • New York • London

Published by Lexington Books
An imprint of The Rowman & Littlefield Publishing Group, Inc.4501 Forbes Boulevard, Suite 200, Lanham, Maryland 20706
www.rowman.com

86-90 Paul Street, London EC2A 4NE

Copyright © 2024 by The Rowman & Littlefield Publishing Group, Inc.

All rights reserved. No part of this book may be reproduced in any form or by any electronic or mechanical means, including information storage and retrieval systems, without written permission from the publisher, except by a reviewer who may quote passages in a review.

Open Access content has been made available under a Creative Commons Attribution-Non Commercial-No Derivatives (CC-BY-NC-ND) license.

British Library Cataloguing in Publication Information Available

Library of Congress Cataloging-in-Publication Data

Names: Kudina, Olya, 1990- author.
Title: Moral hermeneutics and technology : making moral sense through human-technology-world relations / Olya Kudina.
Other titles: Technological mediation of morality
Description: Lanham : Lexington Books, [2024] | Series: Postphenomenology and the philosophy of technology | Originally published as author's dissertation: The technological mediation of morality: value dynamism, and the complex interaction between ethics. | Includes bibliographical references and index. | Summary: "This book considers morality as a dynamic ecosystem that can change in response to its sociomaterial embedding. It particularly explores the role of technology in mediating the meaning of human values and studies the implications of this capacity for the use, design, and governance of technologies"—Provided by publisher.
Identifiers: LCCN 2023036996 (print) | LCCN 2023036997 (ebook) |
  ISBN 9781793651761 (cloth) | ISBN 9781793651778 (epub)
Subjects: LCSH: Technology—Moral and ethical aspects.
Classification: LCC BJ59 .K84 2024 (print) | LCC BJ59 (ebook) | DDC 174/.96—dc23/
  eng/20231004
LC record available at https://lccn.loc.gov/2023036996
LC ebook record available at https://lccn.loc.gov/2023036997

Contents

Acknowledgments	vii
Introduction: Probing the Relation between Technology and Morality	1
1 Morality as an Ecosystem	15
2 Technological Mediation of Morality	31
3 Technological Appropriation and Moral Hermeneutics	49
4 Interpretative Phenomenological Analysis as a Method to Study Moral Hermeneutics	71
5 Hermeneutic Lemniscate as an Encompassing Principle of Moral Sense-Making Mediated by Technologies	103
Conclusion: Reflecting on the Moral Hermeneutics Study from the Perspectives of Technology Design and Governance	121
References	143
Index	157
About the Author	161

Acknowledgments

The research for this book was made possible through funding from the project Value Change, which had received funding from the European Research Council (ERC) under the European Union's Horizon 2020 research and innovation programme under grant agreement No 788321, which also funded the Open Access version of the book; and the Netherlands Organisation for Scientific Research (NWO) under the research program "Theorizing Technological Mediation: Toward an Empirical-Philosophical Theory of Technology" with project number 277-20-006.

Introduction

Probing the Relation between Technology and Morality

ACKNOWLEDGING THE ROLE OF TECHNOLOGIES IN MORALITY

Morality has traditionally been associated with human beings, who independently and reflectively determine the right and wrong courses of action. The rise of technologies in the twentieth century spurred a shift from an independent to an interdependent view of people and technologies, whereby people design new technologies and are profoundly influenced by their own creations. The proliferation of genetically modified food has raised concerns about global justice and responsibility, while the norms of food safety and equitable access to this technology have subsequently undergone review. Assisted reproduction technologies have defied previously existing biological horizons, enabling aged, infertile, or same-sex couples to have children. This has not only redefined what it means to be human but also fostered novel normative expectations regarding procreative rights and liberties. Pervasive internet connectivity and digital technologies implicitly and explicitly weave the canvas of our social and private lives, fostering an expectation of constant availability as well as suggesting who to date, which music to listen to, and what to read. Technologies, thus, while being the fruits of human creativity, manifest not merely as neutral tools but also as productive elements in co-shaping how people perceive the world, each other and themselves.

The role of technologies in forming relations between people and the world is the explicit focus of the technological mediation approach. Positioned within the field of postphenomenology, this approach suggests that technologies are not neutral "objects" in the intentional hands of human "subjects." Rather, they are "mediators" of the relation between people and their environment—technologies mediate human practices and experiences (e.g., Rosenberger & Verbeek, 2015; Verbeek, 2005). Glass office doors enable the expectation of transparency in the professional setting while simultaneously reducing the value of privacy and internalizing a surveilling gaze within employees. Internet-based communication enables the maintenance of long-distance relationships and allows for flexible, permanently connected workplaces, while redefining our moral engagement and responsibility. In this book, I want to venture further into the intricate relation between technologies and morality from the angle of the technological mediation approach. More specifically, I wish to inquire how technologies mediate values and can even change them.

Several authors have written on the ethical dimension of technology, considering how technology embeds values, inspiring human actions and understandings (Parens, 2015); how it fosters moral engagement and relations (Turkle, 2007); and how it can provide moral insights for people (Wallach & Allen, 2009). Moreover, technology can aid in redefining the concepts with which we approach and evaluate it, a phenomenon known as *technomoral change* (Swierstra, 2013).

Magnani (2007) surpasses this idea and suggests that because people and technologies "fold into" each other, producing hybridized entities, morality in a technological world is a dynamic affair. Morality, according to Magnani, is enacted both by people, in changing technological settings and by technologies, to which people delegate their actions and which serve as sources of ethical knowledge. For this reason, people do not *observe* morality when dealing with technologies; rather, the techno-human "folding" *produces* and *reinvents* morality to address changing and challenging situations.

While Magnani's conclusions about the dynamic nature of morality and the active role of technologies within it are persuasive, the rationale behind them is problematic. He argues that because people attribute high value to technologies, people must be considered as "things," as ethical instruments to construct "moral mediators," or "entities we can construct in order to bring about certain ethical effects [be it] as beings, objects or structures" (Magnani, 2007, p. 248). Suggesting the instrumental nature of both people and technologies, the role of technology in how we conduct ethics goes unnoticed. Ultimately, Magnani emphasizes human-technology opposition while downplaying their interrelation. Although the author evaluates technologies ethically, he does not consider the ethics *of* technologies, inviting further considerations in this regard.

The technological mediation approach considers the ethical implications of technologies from the premise of technologies as mediators of human-world relations (e.g., Verbeek, 2011). "If ethics is about the question of 'how to act' and 'how to live,' and technologies help to shape our actions and the ways we live our lives, then technologies are 'actively' taking part in ethics"

(Kudina & Verbeek, 2018). Verbeek (2008), in a study on ultrasound technology, demonstrates that how the image of the fetus appears on the screen has ethical implications for parental deliberation about its future and co-shapes parental responsibilities. By helping to shape moral actions and decisions, technologies mediate morality: prenatal genetic testing mediates moral questions and decisions about childbearing, semi-autonomous robots mediate the moral experiences of war, and CCTV cameras mediate public behavior. Note how, contrary to Magnani's account, when in use, technologies mediate our (moral) engagement with the world not only some, but all of the time.

The postphenomenological contribution to developing the concept of moral mediation is timely and important, for it explicitly reveals how technologies co-shape the moral decisions of people. However, I feel it can be further expanded by considering the relation between technologies and values. My aim in this book is to expand the idea of moral mediation and transcend the postphenomenological scholarship to date. Namely, I wish to demonstrate how technologies mediate not only the moral behavior of people but also the normative frameworks themselves. I believe that more is at stake with the moral mediation of technologies, something that extends beyond the co-shaping of moral intuitions and the decisions of people. I suggest that beyond this, technologies also mediate the infrastructure for moral decision-making, that technologies mediate the meaning of the value frameworks themselves and play a role in value change. The moral mediation account could thus enable the exploration of a continuous development of values related to the sociomaterial contexts in which they are embedded. Throughout this book, I will refer to this phenomenon as *value dynamism* to explore how technologies reveal existing value conceptualizations, thus helping to reaffirm them, shift accents between them, challenge the dominant definitions, and enable new value meanings. Expanding the concept of moral mediation with considerations of value dynamism will dig one layer deeper in the technological mediation approach, clarifying the relation between technologies and values, and expanding its horizons with new inquiries into the dynamics of this relation.

Suggesting that technologies mediate morality does not imply that they define the moral concerns and values for approaching them. Rather, by virtue of their design, foregrounding some options and concealing others, technologies co-shape the (moral) perceptions and actions of people (Ihde, 1993; Verbeek, 2005, 2011). Therefore, technologies themselves do not appear as moral agents; rather, moral agency is distributed among both people and technologies. Moral agency is, as such, a hybrid affair (Verbeek, 2014). Telecare technologies enable physicians to perceive and treat their patients across distances, fostering new configurations of moral engagement and responsibility. Medical imaging technologies guide a physician's interpretation of a patient's health as well as how patients perceive themselves. The ethical implication of this human-technology intertwinement is that technologies also help to shape the moral evaluations and decisions of people (Verbeek 2008, 2011), while, as I wish to add, the ethical frameworks with which we approach technologies co-evolve with these same technologies.

Expanding the idea of moral mediation with that of value dynamism introduces several challenges to the mediation approach as well as to the broader field of the ethics of technology. Concerning the mediation approach, the idea of technologically mediated values highlights the hermeneutic dimension of meaning-making and the interpretation of values in relation to specific technologies. As such, it touches on something broader than the phenomena of value dynamism or value change and concerns the larger process of making moral sense and the interpretative principles that underlie it. I will call this phenomenon *moral hermeneutics* to stress the interactive nature of morality and clarify the interrelation between people and their sociomaterial environment in moral sense-making. As I will argue in this book, particularly in chapter 5, by jointly producing, clarifying, and revising moral precepts that are both produced by their interaction and orientation, human-technology-world relations enact moral meaning. Moral hermeneutics, thus, highlights the meaning-making processes that result in value dynamism and change. With the mediation approach in mind, moral hermeneutics needs to give due conceptual representation of people, technologies, and the sociocultural environment in this regard. To date, however, postphenomenology has emphasized the role of technologies in co-shaping how people relate to the world. When exploring interpretation and meaning-making, it is important to focus not only on specific technologies but also on specific people, with their concerns and sociocultural embedding. Thus, to conceptualize the phenomenon of moral hermeneutics, I must sensitize the mediation approach to the idea of value dynamism and value change and investigate how to study its interpretative dimension conceptually and empirically.

Apart from introducing challenges to the mediation approach, the idea of moral mediation expanded with value dynamism, which complicates the broader practice of the ethics of technology, which is traditionally concerned with anticipating the ethical implications of technologies—a complex and challenging problem. Considering how technologies mediate the meaning of values, which further complicates the practice of ethics. Anticipating the ethical implications of technologies has always been a wicked problem, as articulated by David Collingridge in 1980. On the one hand, we cannot leap over our present shadows to grasp the future implications of technologies while their development trajectory is still flexible. On the other hand, once we know the ethical implications of technologies, they are already deeply entrenched in society and thus very difficult to change. The extent to which I wish to explore the moral mediation of technologies further complicates this dilemma. If we suggest that technologies mediate value frameworks, then how do we still practice the ethics of technology? If the value frameworks that we use to guide the design and evaluation of technologies co-evolve with these same technologies, how do we account for that? Thus, the expansion of the moral mediation idea along the lines of value dynamism presents challenges to the broader field of the ethics of technology.

Thus far, I have established a lacuna in the postphenomenological framework regarding the dynamic nature of values in relation to technologies. I would like to fill this gap by expanding the concept of moral mediation to account for how technologies mediate the meaning of values. Ultimately, I wish to address the question of how technologies mediate values and offer a well-rounded account of moral hermeneutics. This presents, first of all, a theoretical challenge in the technological mediation approach: How do we conceptualize the expanded idea of moral mediation beyond the mediation of moral behaviors and decisions? In parallel, what understandings of values, morality, and ethics would account for the idea of moral hermeneutics? Secondly, a challenge for the practice of ethics ensues from suggesting the dynamic nature of values, whereby the meaning of certain values can change in relation to specific technologies.

Continuing with the practical challenge, suggesting *that* technologies mediate morality does not explain *how* they do so. To address this, I must develop and test a method to empirically study the moral mediation of technologies. While the current section has contextualized the conceptual challenges with regard to value dynamism and the moral mediation approach, the following section focuses on the methodological challenges. It specifically explores how to empirically conduct a philosophical analysis of value dynamism while taking both the methodology and theoretical assumptions seriously.

EMPIRICAL INQUIRY INTO THE MORAL HERMENEUTICS

Verbeek suggests (2015) that people attribute technologies with meaning and importance while appropriating them. For now, I loosely define *appropriation* as a process of taking technologies up, making sense of them, and fitting them into the interpretative frameworks of people. Furthermore, I intuit that moral sensibilities and concerns come to the fore while attributing meaning to a new technology or reinterpreting an old one. If this intuition is correct, appropriation can illustrate how, in an encounter with a technology, pre-reflective values surface, making themselves available for re-articulation and reflection. This book serves as an arena to scrutinize this intuition, attempting to understand how technologies mediate values through the process of appropriation. This places focus on the hermeneutic dimension of technological mediation, on how people interpret technologies and the ensuing mediations. More specifically, I will explore the relation between value dynamism and the appropriation process, attempt to capture the technological mediation of morality, and examine it both empirically and conceptually. This will allow for both the understanding and study of how technologies mediate values and open the door to the phenomenon of value change.

The question itself—how technologies mediate values—requires some clarification, primarily because it draws one's attention to the role of technologies in the mediation process and seems to reduce the visibility of the people who make sense of, use, and reframe these mediations. Additionally, it does not highlight the productive role of the sociocultural environment in giving shape to human-technology relations. Traditions, customs, and power dynamics in society are not just a passive background but help to give course to technological appropriation, at times condemning it from the start, being conducive to it, or conditioning it in other ways. Hence, in providing an encompassing account of moral hermeneutics, it is important to highlight the active role of people and the sociocultural environment in this process.

Throughout the development of the technological mediation approach and the postphenomenological field in general, most of the attention has been on technologies as mediators of human relations with the world. Admittedly, the role of technology has always been correlated with people and their environment. However, since highlighting the status and role of technologies has been the primary concern, it is understandable that human participation in the mediation process, albeit acknowledged, has been less prominent. This must change when exploring how technologies mediate values and how interpretative structures become engaged in this regard. While technologies give shape to moral concerns and, as I will show further in the book, facilitate value dynamism and change, it is people who make sense of technologies and reshape technological mediations. Thus, in pursuing the study of moral hermeneutics, I particularly seek to balance the human-technological-world relation.

The goal that I have identified above, namely, to understand and study moral hermeneutics, and specifically, the technological mediation of values that underlies it, implies a strong empirical component. Just as the phenomenon of value dynamism needs to be philosophically analyzed, it also needs to be captured in one way or another, which would give substance to the conceptual clarifications. How does one empirically identify morality-in-the-making? Earlier, I suggested that values can manifest their malleability in the process of technological appropriation. The preliminary discussions in the previous section hint at the challenges of conceptualizing the relations between technology and morality, particularly regarding the role of technologies in value dynamism and change. The empirical part also deserves close attention, for, albeit promoting the spirit of hands-on research, postphenomenology in general, and the technological mediation approach in particular, is deemed "not empirical enough" (Aagaard et al., 2018, p. xvii).

The Current Status of Empirical Postphenomenology

Postphenomenology studies specific human-technology-world relations, and with that, it combines philosophical analysis and empirical studies. Postphenomenological analysis is empirical in that it studies concrete technologies in embodied human practices. As such, investigations of technological practices go hand-in-hand with philosophical analysis. From its inception, postphenomenology has utilized concrete technological case studies. Ihde, for instance, analyzed, from an auto-ethnographic perspective, cases with acoustic and heart implant technologies, which purportedly made him a cyborg (2002; 2008). He also frequently turned to historic cases of, for instance, cavemen and early painting techniques, as well as reinterpreting history from the viewpoint of material hermeneutics (1990; 2005). Wellner's cell phone case study (2015) became an example within postphenomenology of how technologies can constitute specific conditions of being in the world, that in turn give shape to specific human subjects. In short, postphenomenological studies explicate, through specific cases, how technologies transform human relations with the world, thus co-shaping inclinations and actions.

Through case study analysis, Rosenberger also elicited the political dimension of postphenomenology. He analyzed how technologies in the private and public domains have implications beyond the individual user and always embed the scripts of other stakeholders. For instance, Rosenberger (2009) demonstrated how computers and Internet algorithms enable the formation of filter bubbles. The filter bubbles seduce the user to remain within the space of personal or shared opinions, reducing their exposure to alternative viewpoints, which are essential for an informed perspective. The author (2017) also analyzed the politics of such mundane technologies as benches and trashcans in public places. He highlighted how their design prevents people from sleeping on the street or extracting food from the bins. Such urban technologies draw homeless people away from city centers, creating a perception of having solved the social problem and enhancing a city's image for tourism. In these cases, Rosenberger draws attention to the responsibility of the reflective use of technologies in both the public and private domains and the specific choices of policy-makers.

The 2015 comprehensive volume on postphenomenological research demonstrates the depth and breadth of case study use in postphenomenology (Rosenberger & Verbeek, 2015). The works range from inquiries into what extending the human body through robotic re-embodiment would mean for being human in general and medical practitioners in particular (Besmer, 2015) to analyses of the use of self-tracking technologies when both the "object" and "subject" of tracking are elusive (Van den Eede, 2015). Overall, the volume demonstrates that it is not only possible but also desirable to analyze human-technology relations at the local level of interactions in specific sociocultural conditions. It also illustrates how to philosophically reflect on how technologies co-shape our daily views, preferences, choices, and actions.

However, as Rosenberger and Verbeek (2015) suggest, the use of case studies is left to scholarly interpretation because "There is no strict postphenomenological methodology that scholars should follow" (p. 10). On the one hand, loose methodological guidance invites scholars to experiment with the framework, testing its fitness for a variety of retrospective, current, and future-oriented cases. However, it also runs the risk of neglecting the rigor required of empirical investigations. After all, a careful description and justification of the empirical method enhances the credibility and verifiability of a study and also acquaints the reader with its limitations. To this end, Aagaard et al. (2018) aim to lend postphenomenology scholars a safe footing by suggesting several ways in which a study can be "more empirical." According to Aagaard and colleagues (2018), "Postphenomenologists often base their analyses on texts from science journals and magazines or from their own personal life stories. [. . .] Perhaps [this] has shaped (and restricted) its framework?" (p. xvii). Consequently, the editors of the volume encourage the use of empirical methods from the social sciences to yield new philosophical insights.

This call is most notably represented in several chapters of the volume. For instance, Aagaard (2018) relies on group observation to study the influence of technologies on attention. Secomandi (2018) utilizes interviews, digital ethnography, and group observation to understand the effects of self-tracking technologies on intersubjectivity in human-technology relations. Blond and Schiølin (2018) ethnographically trace the cultural appropriation of the Silbot robot across South Korea, Finland, and Denmark. Finally, Hasse (2018) masterfully utilizes participant observation to suggest how postphenomenology can be elevated from micro-focused, one-person analyses to larger-scale studies. In contrast to these four contributions, the remaining eight chapters predominantly rely on case study analysis, which is very similar to the type presented in Rosenberger and Verbeek (2015). Although the volume purports to focus on the human side of human-technology-world relations, the contributions are structured around specific technological applications, which inadvertently redirects the attention back to technologies. However, as the editors of the volume remark, they do not intend to formulate an exclusive methodology for postphenomenology. Rather, they want to jump-start a conversation regarding the use of empirical methods in this field. With my intention to produce a well-rounded study of how technologies mediate values, I aim to contribute to this empirical call in postphenomenology by substantiating and further extending its methodological horizons.

Empirical Postphenomenology Required to Study Moral Hermeneutics

To understand technologically mediated value dynamism, or morality in the making, requires a strong empirical component that should avoid only being descriptive and "elevat[ing] a single person's self-ethnography to grandiose proportions," as is charged against post-phenomenological case studies (Mol, 2010, p. 254). Equally, as cautioned by Hämäläinen (2016), empirical philosophy should not directly translate empirical insights to philosophical conclusions but rather be reflective and open to the complex translation process that occurs in between. The study of technologically mediated morality must incorporate these methodological suggestions to produce a transparent and critical study, that is empirically grounded but, first and foremost, philosophical.

The version of empirical postphenomenology that I would like to explore must extend beyond the studies presented by Rosenberger and Verbeek (2015) and Aagaard et al. (2018). I intend to focus on the human element in the appropriation process, while accounting for its dynamic technological and sociocultural counterparts. This involves combining empirical insights related to individual experiences, embodied concerns, and personal sociocultural histories while interpreting them through the lens of the technological mediation approach.

Apart from the challenge of incorporating different theoretical and empirical components, the method for studying moral hermeneutics through the prism of human-technology-world relations must be applicable to a wide range of technologies. Throughout the book, I will be drawing on examples of different technologies, regarding both their nature and stages of development: for example, the mixed-reality Google Glass goggles, the voice assistant technologies supported by artificial intelligence such as Amazon's Alexa, the technology for sex selection on a chip, and the COVID-19 tracking apps, among others. All of these technologies present a fruitful but complex background for a study to explore the dynamics of moral hermeneutics.

For instance, the limited introduction of Google Glass in 2013 did not go smoothly. The emergence of "Glass-free zones" and "Glass etiquette" that followed its introduction are examples of the creative appropriation of this technology. These hint at underlying tensions between the values promoted by Glass, such as global connectivity and openness, and those to which people adhere and are not willing to adapt, for instance, privacy. The reintroduction of the device as an enterprise edition in 2019 showcased how the developers tried to incorporate some of these value tensions. The sex selection chip, contrary to Glass, is a technology that is only emerging. It promises to allow parents to choose the sex of their child prior to conception in a safe and cheap manner. Its use for non-medical reasons is particularly interesting, as I will show in the empirical studies I conducted with clinicians and prospective users of the chip. Most legislative systems forbid sex selection for non-medical reasons, but this does not prevent people from escaping national laws and traveling to the countries that do allow this procedure. Such a technological appropriation hints at a conflict between individual and social levels of using sex selection technology (hereafter SST), and it would be interesting to inquire what underlies this conflict.

Even such preliminary reflections on technological practices suggest that understanding how people appropriate a technology and make sense of it in relation to their specific cultural and social settings can serve as an intricate canvas for unraveling moral hermeneutics, i.e., the processes of value identification and potentially revision. In the remainder of the book, I will use case studies in relation to moral hermeneutics to help explore the methodological considerations of studying moral sense-making and value dynamism, as well as examining the mediating role of technology in this process.

THE RESEARCH DIRECTION FOR MORAL HERMENEUTICS

The preliminary theoretical elaborations and intuitions in this chapter allow us to substantiate the direction of this book. A systematic study of the moral mediation of technologies warrants several questions. What exactly is meant by "moral" in this regard? I must also expand the approach of technological mediation in view of the ambition to include value dynamism and the possibility of value change. To highlight the hermeneutic dimension of the moral mediation, one must explore the relation between technological appropriation and the moral sense-making. How should it be conceptualized and studied empirically while bearing in mind the interrelated nature of people, technologies, values, and sociocultural embedding? Next, one must investigate the grounds for a conceptual approach that could integrate the role of technologies in the sense-making process and. if deemed insufficient, offer an alternative drawing on earlier elaborations in the book. Finally, I would like to explore how the moral hermeneutics approaches the building on technological mediation and emphasizes value dynamism, and relates to the fields of ethics and design of technology. Each of the subsequent chapters contributes to solving the conceptual and empirical puzzle of making moral sense in, of, and through human-technology-world relations.

Chapter 1 clarifies what the term "moral" means in relation to the ideas of moral hermeneutics and the technological mediation approach. It lays the foundation to expand the technological mediation approach with value dynamism and change and highlights the importance of outlining a comprehensive principle of moral sense-making. In this, I rely on the pragmatist origins of postphenomenology, specifically, the works of Dewey. Turning to pragmatism allows me to present a theory of values that embeds their relational nature within the sociomaterial environment. To unpack the "moral" part of the moral hermeneutics in relation to technologies, I particularly clarify which concepts of morality, ethics, and values it entails. I rely on the pragmatist origins of the mediation approach to emphasize the dynamic model of sociomaterial practices and to position values as both enabled by and conditioning these practices. Ultimately, I define the perspective of the mediation approach on values as relational, dynamic, and flexible regarding the human-technology-world practices in which they are embedded. This lays the ground to explore the hermeneutic dimension in the mediation approach regarding how values can undergo conceptualization, reinterpretation, and change.

In chapter 2, I trace how the question of technology and values is represented in the field of ethics of technology, in the mediation approach, and beyond. To do this, I first explore how the technological mediation approach has to date considered the ethical dimension of technologies, particularly concerning co-shaping moral perceptions and actions. I then focus on the technomoral change approach of Swierstra, which suggests that technologies induce moral change. Finally, I draw on the sociotechnical experimentation approach of Van de Poel and his recent work on value change. The latter is particularly interesting because it has developed ways for exploring different mechanisms through which technologies enable value change, such as value dynamism, value adaption, and value emergence. Discussing the relation between these three approaches allows me to delineate the goals of technological mediation with regards to moral hermeneutics and suggest how its scope should be expanded to include the phenomena of value dynamism and the overall hermeneutic dimension of moral sense-making. To do this, I once again turn to pragmatism in order to sensitize the mediation approach to the ideas of value dynamism and change.

Chapter 3 develops the line of appropriation to understand how people make sense of technologies in a projective and practical manner and how this relates to the technological mediation of morality. I first clarify the concept of appropriation in the mediation approach and compare it to the existing concept in the domestication studies. I then present a brief ethnographic study that explores the appropriation of Google Glass on YouTube and presents a preliminary empirical verification of the idea of technological mediation expanded with value dynamism. Aided by online comment analysis, I tentatively explore the relation of Google Glass to the value of privacy. The study demonstrates how various dimensions of the value of privacy resurface in relation to Google Glass and the specific practices that it enables. The preliminary study functions not only to provide answers but also to further flesh out the research lines. The function of the case study in this chapter is thus threefold: first, to illustrate how appropriation can be a first step in exploring moral hermeneutics; second, to demonstrate the urgency for an empirical and conceptual inquiry in the different modes of moral sense-making that would accommodate the human, technological, and sociocultural counterparts, and different forms of value dynamism; and third, to give an idea about potential theoretical and empirical challenges in studying moral hermeneutics before plunging into deeper theoretical and methodological investigations.

In chapter 4, I develop a methodological framework to empirically study the process of moral hermeneutics that would be able to capture technological appropriation and uncover the dynamics of values. I explore the method of Interpretative Phenomenological Analysis (IPA) and suggest that it fits the sense-making goals of the study because of its focus on situated microperspectives and on the philosophical principles of circular interpretation as developed by Gadamer. I next test out the IPA method combined with the theoretical ideas developed in the earlier chapters to the case study of a technology in the making, the prenatal sex selection chip. The sex selection chip, contrary to Google Glass, primarily exists in the form of promises, concerns, fears, and ethical debates. The moral hermeneutics here thus concerns the projective modes of appropriation and shows how already at this stage of sense-making, technology mediates the normative assumptions of people related to good parenthood, liberalism, naturalness, and others. I end the chapter with preliminary conclusions on examining moral hermeneutics through the empirical lens of IPA, suggesting that, albeit time-consuming and challenging, it can be helpful for the more informed use and decision-making on technologies.

In chapter 5, I develop a hermeneutic lemniscate as a principle of technologically mediated interpretation that helps us to understand and navigate the dynamics of moral hermeneutics. I refer to such an integrated process of interpretation as a lemniscate because it resembles a twisted figure-eight shaped curve (∞), consisting of three linking, interrelated components: humans, technology, and the sociocultural world. With the lemniscate, I extend the hermeneutic dimension of postphenomenology to account for the missing mechanism of circularity between people and the mediated world. To expand postphenomenology, I build upon Gadamer's circular principle of interpretation. In turn, I argue that Gadamer's account misses the mediating part of technologies in the hermeneutic circle and thus turn to the material hermeneutics of Ihde that explicitly accounts for it. Combining Gadamer's and Ihde's accounts allows me to produce an encompassing account of technologically mediated moral sense-making, the hermeneutic lemniscate. The first line of the lemniscate, from human through technology to the world, designates the appropriation process of a technology. The second line of the lemniscate, from the mediated world back to the person, designates the process of subject constitution and the update of interpretative references. Values, according to the pragmatist definition in chapter 1, are both enabled and reconfigured in the human-technology-world interrelation. Together, the appropriation and subject constitution lines of the lemniscate mirror the dynamic, open, and fluid process of moral hermeneutics. Analyzing through the prism of lemniscate how people appropriate technologies, how the mediated sociocultural world shapes specific subjects and feeds back into the interpretative schemas of people, allows me to see how the normative concerns surface when considering specific technologies and how the human-technology-world relations offer spaces for re-articulating values. Throughout the chapter, I refer to the case study of AI-enabled voice assistants such as Apple's Siri and Google's Assistant for illustration and analysis.

Finally, the conclusion reflects on the conceptual and empirical findings of the moral mediation approach as expanded from the perspective of moral hermeneutics. Here, I reintroduce the ethical variant of the Collingridge dilemma, briefly explained in the introduction, and relate the mediation approach to the existing philosophical approaches of Swierstra and Van de Poel discussed in chapter 2. Next, I relate the book's findings back to the field of technology design, namely the approach of Value Sensitive Design. Throughout the chapter, I also discuss the potential challenges of using the perspective of moral hermeneutics and sketch further research avenues.

Chapter 1

Morality as an Ecosystem

EXPLORING THE "MORAL" PART IN THE MORAL MEDIATION ACCOUNT

In the previous chapter, I set out to explain how technologies mediate values. Doing so requires expanding the moral mediation account with ideas about value dynamism. This chapter explores the question of how to theorize the phenomenon of moral mediation so that it accommodates such ideas. Postphenomenology scholars emphasize the ethical dimension of technologies by highlighting the link between technological design and the ensuing perceptions and behaviors of people (e.g., Van den Eede, 2015; Rosenberger, 2017; Wellner, 2015). The moral significance of technologies received explicit attention in Verbeek's (2011) moral mediation account, which argues that by co-shaping the moral intuitions and perceptions of people, situations of moral choice and habits, technologies mediate morality. However, as I suggest in the introduction and will demonstrate further in the book, co-shaping the moral perceptions and actions of people is only one facet of the technological mediation of morality. Technologies also mediate value frameworks, whereby values both guide people in decision-making (about technologies) and appear to be mediated by these same technologies. This chapter builds toward a systematic theoretical understanding of moral mediation as expanded with value dynamism. To this end, it defines what the terms "moral" in the technological mediation account means, and will clarify the background assumptions and conceptual frameworks underpinning the moral mediation of technologies.

Returning to the pragmatist origins of postphenomenology, specifically the works of Dewey, aids me in my conceptual pursuits. In this chapter, a pragmatism lens helps to explain the multidimensional nature of values, their openness to revision, and their intrinsic relation to sociomaterial practices.

In the following chapter, the turn to pragmatism will allow me to expand the moral mediation account with the ideas of value dynamism by deepening its understanding of interrelational ontology. As such, a pragmatism angle can help to explain how the mediation approach is a special bridge between technologies and values. Understanding the moral mediation of technologies requires embracing the pragmatist origins of mediation more explicitly than postphenomenology currently does. The steps above will ultimately allow me to produce a comprehensive and encompassing understanding of the moral mediation of technologies expanded with value dynamism and build toward an overarching idea of moral hermeneutics.

In what follows, I explore the "moral" part of the moral mediation account. What is meant by "values" in the technological mediation of morality? How does the mediation perspective relate to the formal theories of values? How can the pragmatism of Dewey help me to solidify a perspective on values in relation to my previous questions? To be clear, asking these questions does not presuppose an analytic exercise to dissect values as to their origin, nature, and character, arguing for one position or another. Nonetheless, the inquiry in this chapter bears some resemblance to an analytic exercise, necessary to elucidate the position and ambition of the technological mediation approach as I understand them. Turning to Dewey and exploring formal value theories allows me to identify values in the technological mediation approach as not fixed but dynamic, open to revision and sensitive to the sociomaterial context. An example of COVID-19 tracking apps in the following section can help to begin analyzing the dynamic nature of values.

A PRELIMINARY PRACTICE-BASED EXAMINATION OF VALUES: A CASE OF COVID-19 TRACKING APPS

Before I proceed to the reflections of values and morality against ethical theories, I would like to briefly discuss an intimate interrelation between values and technologies with an example of COVID-19 tracking apps. At the onset of the pandemic in early 2020, national governments across the world aimed to technologically facilitate the reduction of virus exposure by tracing the carriers and alerting the people who might have been in close proximity. This effort resulted in the development of multiple mobile applications for contact tracing, often known as Corona apps. However, their introduction into society did not go smoothly. Faced with the persistent moral uncertainty accompanying the pandemic, people did not know how to position the apps alongside the existing measures of wearing masks, social distancing, and working from home.

Additionally, there were a lot of contradicting media messages as to the technological capabilities of the apps (Kudina, 2021a). Because most governments opted for developing separate national applications, there was a lot of difference as to what they could and could not do. For instance, while some contact tracing apps relied on precise GPS location tracking of the users (e.g., Iceland, Israel, Iran, etc.), many others used GPS signals only for triangulating the data streams and could not store or share the exact whereabouts of the app users (e.g., most EU apps). While some countries enforced a mandatory use of the COVID-19 tracking apps (e.g., China, South Korea), others highlighted their voluntary nature (e.g., the USA). What further accompanied the uncertainty surrounding the apps was the fact that the app development most of the time, relied on a corporate collaboration with Google and Apple that provided a notification exposure infrastructure for the tracking apps. While the rest of the app could be built in an open-source manner, the key functioning of the app remained proprietary and secret, even though the companies provided assurances that the data they receive from the apps would not be used for commercial purposes. Together, this resulted in a public discourse about the tracking apps that pitted privacy and solidarity against each other. In the absence of broader awareness campaigns and technical explanation, COVID-19 tracking apps appear as either the surveillance dystopia propelled by the government engaging with private corporations or a collective responsibility to provide a united response to the pandemic, privacy risks notwithstanding (e.g., Dodd, 2020, Lovett, 2020, Siffels, 2020).

The contact-tracing apps prompted the emergence of new social norms and a reconfiguration of the existing ones. Conflicting media reports, corporate collaboration, and the absence of clear early on messages about what the apps can and cannot do resulted in understandable privacy concerns that drove the non-adoption of the apps (Jansen-Kosterink et al., 2021). Further, there was a worry about the apps being a technological fix to the pandemic, creating a sense of false security and prompting people to disregard other social measures based on the fact that they are not receiving app alerts (Williams, 2020). Meanwhile, most governments stressed the voluntary nature of the apps while at the same time motivating people to use them by highlighting the individual responsibility to help others and the value of solidarity. This raised some feelings of indirect coercion, a possibility for discrimination, and social punishment if they chose not to use the tracking apps (Siffels, 2021). Finally, some people were concerned with the lack of government strategies regarding the corporate collaboration with Google and Apple, underscoring the power of the private companies to determine the public health concerns and the potential long-term costs of such collaborations (Sharon, 2021). Overall, the introduction of COVID-19 tracking apps was accompanied by moral uncertainty, diverging interpretations of individual responsibility, trust, public health, privacy, and perceived conflict between the individual and the collective.

COVID-19 tracking apps present a case in point of the technological mediation of morality. The new social situation of the pandemic and the uncertainty surrounding the apps invited people to reveal their usually dormant moral views and ideas, reflecting on their own actions or those of others and also anticipating potential new situations with such apps. The illustration above also suggests how the usually abstract moral values (such as privacy, responsibility, respect, etc.) that guide and inform us in our daily behavior require contextualization and substantiation in view of new technologies. COVID-19 tracking apps introduced several new complex issues for people to consider. The case also indicates how multiple practical and moral concerns materialize in discussion about this new technology. The tracking apps appear to be far from neutral in moral deliberations and act as a catalyst for ethical reflection and decision-making. While some people refused to use them altogether, others embraced them, and some developed selective use, e.g., turning the app on the go and turning it off when arriving at social gatherings, so as to prevent potential multiple notifications and battery drainage from the app (Kudina, 2021a). In short, this short case study examination suggests how technology can invite people to foreground their normative assumptions and review or update them in light of the novel situation and material context that a new technology creates.

Technology is far from neutral; it mediates both the moral behaviors and the moral infrastructure for decision-making, the conceptions of the values themselves. With regard to privacy and solidarity, the tracking apps case illustrates that the concrete substantiation of values in a practical sociomaterial situation is often conflicting and manifests in more than one way, although the abstract ideal can be the same (Mol, 2002). This means that values require a reflection process to clarify their meaning and determine the course of action (Swierstra et al., 2009). Values appear not as isolated from the surrounding social and material situation but as both embedded in it and co-constituting it. In other words, values, through this short study, emerge as relational to the social and material counterparts of the given practice (Verbeek, 2011). Next, I wish to substantiate the definition of values in the technological mediation approach, considering the mediating role of technologies in co-shaping their formulation and interpretation.

A POSTPHENOMENOLOGICAL TAKE ON VALUES

In the moral mediation account of Verbeek (2011), in exploring how technologies mediate moral perceptions and actions, values were regarded as a technological impact, a consequence of technological mediation. Their role as moral infrastructure entangled with human-world experiences, both conditioning and being a product of technological mediation, was not considered. Both the brief exploration of the COVID-19 tracking apps and the preliminary conceptual reflections above suggest values through a lens of sociomaterial practices. The turn to such practices has been recently gaining strength in Science and Technology Studies (STS) and the philosophy of technology (e.g., Mol, 2002; Suchman, 2007; Latour, 2005), because it emphasizes "the constitutive entanglement of the social and the material in everyday [. . .] life" (Orlikowski, 2007, p. 1438). The practice lens offers the opportunity to view people as inextricably related to technologies in the practices that the two enact together in the larger social and cultural environment. Moreover, the practice approach

> encourages us to regard the ethical problem as the question of creating and taking care of social routines, not as a question of the just, but of the "good" life as it is expressed in certain body/understanding/things complexes (Reckwitz, 2002, p. 259).

Thus, adopting the approach of sociomaterial practices invites the expansion of ethics from traditional interhuman relations to include relations with technologies (ibid.).

A practice lens is not alien to the technological mediation approach, which positions technologies within concrete human-world relations and practices. I suggest the application of a lens of sociomaterial practices to clarify the role and character of values in mediation. A practice-based approach belongs to the theoretical foundation of postphenomenology, drawing its origins from pragmatism. Making the pragmatist roots in the mediation approach more explicit can help to clarify the nature and role of values with regard to human-technology-world relations.

Ihde (2009) identifies postphenomenology as building on pragmatism (primarily after Dewey), phenomenology, and the philosophy of technology. Pragmatism grants postphenomenology a reliance on the dynamic and interrelated environmental model of human experiences. According to Ihde, the pragmatism of Dewey avoids the subject-object distinction by framing human experiences as relational environments embedded in concrete material and social settings (Ihde, 2009). Ihde compares Dewey's pragmatism to Husserl's phenomenology and contends that, while Husserl adopted the early modern "subject–object," "internal–external" vocabulary,[1] Dewey avoided it by beginning with embedded practices and experiences (2009, pp. 9–10). According to Ihde, pragmatism relieves the phenomenology of the subject-object vocabulary because its reliance on experience "short-circuits the 'subject/object' detour derived from Descartes—or [. . .] Locke—and points much more directly to something like a lifeworld analysis" (p. 11). Ihde considers it "a way to avoid the problems and misunderstandings of phenomenology as a subjectivist philosophy, sometimes taken as antiscientific, locked into idealism or solipsism" (2009, p. 23).

Another important feature that Ihde identified in Dewey's pragmatism and integrated into postphenomenology is "interrelational ontology,"[2] whereby "the human experiencer is to be found ontologically related to an environment or a world, but the interrelation is such that both are transformed within this relationality" (2009, p. 23). In relation to values, interrelational ontology could explain how values respond to fluid sociomaterial environments and could themselves be dynamic and flexible, as in the Glass and privacy example. I suggest more closely examining Dewey's environmental model of human experiences in relation to values. A pragmatist stance toward values as *values-in-practices* could account for their dynamic, relational nature within the human–technology–world ensemble and hint at how technologies mediate values.

According to what I have suggested above, preliminary values emerge as relational, dynamic, and embedded in lived practices. To scrutinize my hypothesis, I next position these preliminary value considerations against existing formal ethical theories on values and the pragmatist scholarship of Dewey. I explore how different value ideas correspond to the pillars of technological mediation as a profoundly relational, phenomenological, and experience-bound approach to empirical philosophy.

EXAMINING VALUES THROUGH FORMAL THEORIES

The normative component in the technological mediation of morality is the values that technologies mediate. The concept of value has formed the foundation of many diverging accounts throughout the history of philosophy. Not surprisingly, although the formal aspects of value suggest some unity in grasping its meaning, concrete substantiation varies greatly.

Dictionaries and formal theories of value provide a historical consensus point of departure for further elaboration on values. Definitions of value greatly vary per discipline. I consider those definitions of value primarily related to the fields of philosophy and sociology, discarding the economic definitions of value related to price and monetary benefit. According to the *Encyclopedia of Philosophy*, values denote the worth of something and that which it *"ought to be"* (Frankena, 1967, p. 637). The *Oxford Dictionary of Philosophy* provides a broader definition of value as an aspect of a phenomenon included in decision-making, "a *consideration in influencing choice* and guiding oneself and others" (Blackburn, 1996, p. 390). These definitions concern value as a noun, something to which one can apply qualitative terms, such as *"good, desirable or worthwhile,"* and that covers "all kinds of rightness, obligation, virtue, beauty, truth, and holiness" (Frankena, 1967, p. 637). Value as a verb concerns the practice of *perceiving and attributing value*, which frequently—but not necessarily—involves comparison and reflection, with the terms "to value," "valuing," and "valuation" reflecting this meaning (ibid.).

The *Oxford Dictionary of Sociology* adds to these definitions a consideration of the nature of value, defining it as "strong, semi-permanent, underlying, and sometimes inexplicit dispositions" (Scott and Marshall, 2009). Such a sociological definition invites considering another discussion regarding values, namely, the nature of their relation to that which is valued. Some consider that values are *intrinsic* properties that showcase the inherent worth of a phenomenon and thereby are independent from the likings, beliefs, and preferences of those engaged in valuation. Plato is one of the notable philosophers who endorses the view of values as intrinsic goodness. Such philosophers are sometimes called intuitionists to denote that appreciating an intrinsic value requires some sort of intuition, "a special sort of awareness or though process to detect it" ("Value," 2005, p. 941). By this token, value is a non-derivative property, "an indefinable non-natural or nonempirical quality or property different from all other descriptive or factual ones" (Frankena, 1967, p. 640). In connection to this, value is considered intrinsic, nonempirical property of a phenomenon, linked to an *objective* account of values that is independent of human perception and is thus a matter of fact.

Others posit that values are *relational* properties of a phenomenon, whereby the phenomenon derives its value from being valuable for achieving certain ends, or valuable *as* something *for* someone. Alternatively, but also within the relational account, values may not be perceived as properties of a phenomenon but as "a matter of the loving regard we pay to things" ("Value," 2005, p. 941). Aristotle, a notable representative of the relational account of values, considered the phenomenon of value only inasmuch as it is an object of desire or interest (Frankena, 1967, p. 638). As opposed to the objective account, which regards values as a matter of fact, a subjective account may be "a product of the mind of the judging subject" ("Value, aesthetic," 2005, p. 941). Paraphrasing a famous saying, "Value is in the eyes of the beholder." However, it is misleading to dismiss the relational account of values based on its assumed subjective nature. As the saying indicates, to be of value, a phenomenon must both possess worth defined as valuable and "the eye of the beholder" that brings it into existence.

Other positions attempt to present a middle ground between the objectivist and subjectivist perspectives regarding the source of value, for instance, the *rationalist* (Van de Poel, 2009a), who positions the origin of value in *human rationality*. While the charge against subjectivists is that they may confuse value with human preference and desire, the charge against objectivists is that they entirely divorce value from it. In this sense, the rationalist position is the middle ground between the two:

> It restores the connection between human desires and values, which is lost in objectivism but strives to avoid confusing value with preference by claiming that things are valuable not just because people prefer them but because rational beings have sufficient practical reason to pursue them (Van de Poel, 2009a, p. 976).

However, if one more closely examines the assumptions of the rationalists, their inevitable proximity to the objectivist perspective becomes evident. By attributing primacy to human rationality in being able to identify and act on certain values, the claim is essentially that all rational human beings can identify and choose what is valuable. But what is it that makes the phenomena stand out to them as valuable? Something must exist in the phenomena themselves that all critical, reasonable people perceive as valuable. This position closely approaches the objectivist's thesis concerning the inherently valuable properties of the phenomena.

Van de Poel (2009b), acknowledging the shortcomings of the three positions above, suggests a "mildly rationalist" perspective on values that avoids the extremes of both objectivism and subjectivism. This position is only mildly rational in suggesting that "values should be distinguished from preferences but not completely divorced from human desire, interest, interpretation and meaning-giving" (p. 976); at the same time, it does not adhere to one specific view on the source of values.

Another philosophical position that attempts to reason with the opposing value dichotomies of objective-rational and subjective-empirical is the experimental empiricism of John Dewey (1930). Rather than focusing on the question of value definition, Dewey is instead concerned with *value formation*. Dewey, who famously declared that "[t]he separation of warm emotion and cool intelligence is the great moral tragedy" (1922, p. 258) and attempted to reconcile the two in his moral philosophy, was greatly inspired by the thought of Aristotle. On the one hand, he vehemently opposed the conception of values as eternal, final, universal, and rational, which, according to Dewey, would detach them from the conflicting lived experiences of people and defy the idea of values as orientation in action. On the other hand, he also criticized the empirical approach to values because it denigrates the status of values to mere preferences and subjective enjoyments. According to Dewey,

> Without the introduction of operational thinking, we oscillate between a theory that, in order to save the objectivity of judgements of values, isolates them from

experience and nature, and a theory that, in order to save their concrete and human significance, reduces them to mere statements about our own feelings (Dewey, 1930, p. 263).

Contrary to the above positions, the nature and source of values appear to Dewey as "experimental empiricism." According to this approach, value derives its guiding character and meaning from the connection to particular situations and the reflective judgment upon them. This is what makes values "the fruit of intelligently directed activity" (1930, p. 272). For Dewey, values are "based upon concern with facts and deriving guidance from knowledge of them" (1922, p. 12). As practical judgments and orientations in concrete experiences and practices, they are both empirical and regulative (1930, pp. 256–269).

What particularly distinguishes Dewey's account of values is its inherent intertwinement of values with lived practices. To better understand the process of value formation, Dewey suggests moving away from a theory of values and toward a *theory of valuation* (1939), which implies studying concrete practical situations that specify a given value. Because of the integral intertwinement of values and the practices in which they are embedded, Dewey defines values not as normative ends but as "ends-in-view" (1922), which are not objective and universal fixed finalities, closed off or isolated. Rather, values as ends-in-view denote hypotheses and aims that both guide the present action and are tested by it, adjusting in their dynamism to the exigencies of reflection, choice, and effort. A conception of values as "ends-in-view," or "endless ends" (1922, p. 232), suggests that every new practical constellation can enable new ends as both means and regulative guidelines.

In addition, a practice-oriented take on values allows Dewey to account for both the universal and particularistic conceptions of value. Values as "ends-in-view" assume a default general position from which reflection can proceed, contextualizing it per given situation. Dewey emphasizes this hermeneutic relation between a universal and a contextualized meaning of values with the example of a patient visiting a doctor. An abstract idea of health always guides the doctor during examination, but the patient's medical history, condition, and supporting factors determine health as the end-in-view:

> [The doctor] forms his general idea of health as an end and a good (value) for the patient on the ground of what his techniques of examination have shown to be the troubles from which patients suffer and the means by which they are overcome. There is no need to deny that a general and abstract conception of health finally develops. But it is the outcome of a great number of definite, empirical inquiries, not an a priori preconditioning "standard" for carrying on inquiries (Dewey, 1939, p. 46).

Defining what health means for a particular patient requires possessing an abstract notion of health, just as understanding multiple dimensions of value in a specific setting requires an abstract formulation of that value. Such abstract assumptions are often widespread, lending themselves to the process of valuation as useful starting points. However, they point only to what may be at stake because new concerns and practical exigencies necessarily arise in the course of their valuation as new ends-in-view.

In summary, the dictionary definitions and moral theories of value can be useful to illustrate how the formal interpretations of value have developed over time, conflicted with, and aligned with each other. I think that both the dictionaries and value theories provide the time- and experience-proven faceting of the gems they present values to be. However, to get to the messy, unpolished core of values, one must cut and facet the stone oneself. This is possible only when implementing the value definitions and verifying them within practical situations; otherwise, the dictionary definitions and moral theories remain detached from reality. They can theoretically guide one in abstract terms but can only rarely be applied in dilemmatic situations. The pragmatist definition of values as lived in human experiences, shaping and co-shaped by the sociomaterial surroundings, is better suited to deal with the complexity of daily life. How would it correlate with the mediation approach, particularly regarding explaining how technologies mediate values? To answer this question, I further elaborate upon the scholarship of Dewey, which also allows me to substantiate the larger philosophical ambitions of the technological mediation approach.

SOLIDIFYING VALUE PERSPECTIVES THROUGH PRAGMATISM

Discussing the nature and source of values illuminates the variety and complexity of perspectives on the issue. I do not intend to define which position is the right one, but instead, I wish to correlate existing perspectives on values with the approach of technological mediation. When one discusses technology mediating values, what is meant by "value?" Furthermore, what type of moral philosophy would this imply? I suggest that the pragmatist philosophy of Dewey is very similar in spirit and method to the technological mediation approach, and as such, examining it more closely can help to expand its moral dimension.

Relationality is a core principle upon which the mediation approach and pragmatism coincide. Dewey posits that a phenomenon inevitably possesses intrinsic irreducible qualities that condition its existence. However, conceived only in these terms, it is subject only to existence but not presence or being. *Being is a relational experience* that necessitates knowing, denoting, describing and feeling. Just as Heidegger, Merleau-Ponty, and Gadamer later asserted from the phenomenological standpoint, Dewey, in his pragmatic inquiry in the early 1920s, suggested that all activity acquires meaning in relation to the surrounding counterparts, be it people or specific technologies. According to him,

> experiences of contact with objects and their qualities give meaning, character, to an otherwise fluid, unconscious activity. We find out what seeing means by the objects that are seen. They constitute the significance of visual activity that would otherwise remain blank. [. . .] [Activity] acquires a content or [. . .] meanings only in static termini, what it comes to rest in [. . .] [T]he object is that which objects (Dewey, 1922, p. 191).

Ihde integrated the idea of interrelational ontology from pragmatism to present it in the mediation approach, with regard to the co-constitution of people and technologies (2009, pp. 10–11, 23). I suggest further extending it to values in human-technology-world relations.

Dewey can help clarify the specific subjectivity implied in a relational conception of values (1929; 1930). Commenting on the status of values in nature, he distinguishes between the immediacy of existence and the actual presence of phenomena (such as values). In relation to the discussion of values above, it mirrors the debate between values conceived as intrinsic or relational.

> [The] irreducible, infinitely plural, undefinable and indescribable qualities [a] thing must *have* in order to be, and in order to be capable of becoming the subject of relations and a theme of discourse. [. . .] Description [. . .] [is] index to a starting point and road which if taken may lead to a direct and ineffable presence. To the empirical thinker, [. . .] nature has its finalities as well as its relationships (Dewey, 1929, pp. 85–86, original emphasis).

In the same manner, Dewey positions values in relation to environmental factors by suggesting that "until the integrity of morals with human nature and of both with the environment is recognized, we shall be deprived of the aid of past experience to cope with the most acute and deep problems of life" (1922, pp. 12–13).

I find Dewey's ideas important because they can help the moral mediation account to avoid the potential relativism charge. As Pitt notes, as soon as a philosophical account endorses ideas of value dynamism and change, a "standard worry" emerges: "Aren't we doomed to relativism?" (2014, p. 92). One could argue that suggesting that technologies mediate values implies that values are subject to individual preferential interpretation or that both everything and nothing is of value. From there, it is only one step to nihilistic ideas about possessing no morality at all, no moral rules or contracts. I believe these relativist worries should not be easily dismissed, but at the same time, they are not warranted when value dynamism is concerned. Following the pragmatist spirit, it is shortsighted to dismiss the relational conception of values as relativistic because this presupposes a qualitative empirical basis. Dismissing the relational perspective on values and the ensuing value dynamism would mean rising above the lived and felt world of uncertainty and conflict and reducing values to the isolated space of the universal and objective, but at the same time, the indescribable and unknowable (Dewey, 1930, p. 278). Perceived, felt, and lived values manifest in the relation of people with their environment, both cultural and material. Different sociomaterial constellations can enact different manifestations of the same value. An expanded moral mediation approach does suggest that values are embedded in sociomaterial practices, which explains their dynamic and interrelated nature. However, it does not introduce an "anything goes" approach regarding values but rather emphasizes a developmental aspect of values.

In a similar vein, I consider the mediation approach to assume a bridging position between objectivist and subjectivist accounts of values. What sets this approach apart from Dewey's pragmatism is its explicit foregrounding of the material dimension of mediation. On the one hand, the mediation approach understands values as relational properties, enabled and manifested in the experiences of people in relation to their sociomaterial environment. On the other hand, it avoids the subjectivist-objectivist divide by acknowledging that values are neither mere subjective preferences of people nor simply static matters of fact. Rather, in the mediation approach, values appear as a lively and complex matter of concern, interest, and care, as opposed to a matter of fact (Latour, 2004, 2008; de la Bellacasa, 2011; Dussauge et al., 2015). Explicitly acknowledging the role of technologies in co-shaping values and moral concerns organically expands the relational considerations of values, introduced by Aristotle and further developed by Dewey, and adds a novel and specific concern regarding the role of technologies in enacting and realizing values (Verbeek, 2011). This offers new avenues for the relational approach with regard to technologies.

By the same token, the mediation approach also opens new modes of inquiry for itself, suggesting that it combines empirical knowledge of technologically mediated human practices with systematic philosophical analysis (Verbeek, 2011). In his time, Dewey a propelled moral philosophy that does not shy away from the messy lived human experiences but incorporates them as the basis for philosophical inquiry. Verbeek challenged philosophical practice by suggesting that it should account for the mediating role of technologies in moral affairs. Understanding Dewey's scholarship on reflective inquiry can be instructive for understanding the challenge Verbeek introduces.

For Dewey, the empirical world of qualitative values and rational knowledge need each other to live intelligent, reflective lives: "Our affections, when they are enlightened by understanding, are organs by which we enter into the meaning of the natural world" (1930, p. 282). An organic intertwinement of values and knowledge, conceived as critical engagement, lies at the core of Dewey's moral philosophy. The critical engagement model very closely resembles how I view the philosophical position and ambition of the technological mediation approach. This approach expands the relation between value and critical reflection by acknowledging the moral significance of technologies and by urging that it be accounted for.

According to Dewey, the task of philosophy consists of deliberate critical inquiry into the nature of human experience as entangled with values. In other words, its goal is to formulate the basis for an informed choice for further actions:

> Its primary concern is to clarify, liberate and extend the goods which inhere in the naturally generated functions of experience. [. . .] This does not mean [philosophy's] bearing upon *the* good, as something itself attainted and formulated in philosophy. For as philosophy has no private score of knowledge or of methods for attaining truth, so it has no private access to good (Dewey, 1929, pp. 407–408, original emphasis).

Dewey is very clear that his moral philosophy bears no prescriptive features or grounds in any traditional normative accounts, be it in the utilitarian pursuit of the greatest happiness or in a deontological reliance on human reason, as he believes that this would contradict the sole purpose of reflection and critical inquiry into the relational nature of moral pursuits. For Dewey, philosophy is a critical messenger, not a messiah:

> The need for reflective morality and moral theories grows out of conflict between ends, responsibilities, rights, and duties [and] defines the service which moral theory may render [. . .]. [I]t does not offer a table of commandments in a catechism [. . .]. It can render personal choice more intelligent, but it cannot take the place of personal decision, which must be made in every case of moral perplexity. [. . .] [T]he attempt to set up ready-made conclusions contradicts the very nature of reflective morality (Dewey and Tufts, 1908/1932, pp. 175–176).

Dewey's idea of moral philosophy to aid in informed decisions but not prescribe the decisions themselves closely resembles the nature of the technological mediation approach, as suggested by Verbeek (2011). As demonstrated above, the starting point of Dewey's philosophical inquiry was to break the walls artificially constructed in philosophy that separate people from their environment. The mediation approach begins with an emphasis on technologies as a counterpart co-shaping people and the world in which they realize themselves: "Whoever fails to appreciate how technology and humanity are interwoven with each other loses the possibility of taking responsibility for the quality of this interweaving" (Verbeek, 2011, p. 155). Verbeek aims to account for such a co-shaping ensemble of people, technologies, and the world in both theory and practice. He consequently reshapes the famous virtue ethics question, with technological mediation in mind, to ask "How to live a good life with technologies?" (2011, pp. 156–158).

Verbeek, as does Dewey, opposes the role of philosophy and ethics as being prescriptive and defining ultimate moral positions in dealing with technologies. He argues that "Instead of making ethics a border guard that decides to what extent technological objects may be allowed to enter the world of human subject, ethics should be directed toward the quality of interaction between humans and technology" (Verbeek, 2011, p. 156). More specifically, in the context of the technological mediation of morality, the goal is also to show how the moral standards, principles, and values with which we approach technologies co-evolve in relation to these same technologies. Ethics, understood as a practical moral philosophy, appears here as a critical inquiry, a reflection on the quality of human-technology intertwinement in the world, and specifically about the moral significance of the technologies in it. Ethics presents information for decision-making to the users, non-users, designers, and policy-makers. However, such ethics will not decide for the actors concerned but instead accompany them in the process of decision-making.

To address the mediating presence of technologies in our lives and their mediating capacity for human values, the technological mediation approach aims to reconcile the empirical and normative focus in the philosophy of technology. The following quote illuminates Verbeek's specific viewpoint on this matter:

> Accompanying technological developments requires engagement with designers and users, identifying points of application for moral reflection, and anticipating the social implications of technologies-in-design. Rather than placing itself *outside* the realm of technology, an ethics of accompaniment will engage directly with technological developments and their social embedding. Its primary task is to equip users and designers with adequate frameworks to understand, anticipate, and assess the quality of social and cultural impacts of technologies. This type of ethics therefore requires an integration of the empirical turn and the ethical turn (Verbeek, 2011, p. 165, original emphasis).

Thus, echoing Dewey's urgency regarding grounding philosophical inquiry in concrete experiences, Verbeek suggests that philosophically scrutinizing human-technology-world relations is impossible without exploring the messy empirical realities of human practices. Dewey, however, focused on the single citizen empowered by the critical inquiry. Verbeek, conversely, considers a broad range of stakeholders who can and should adopt a reflective stance on technologies in both use and design. Users, in their daily interactions with technologies, directly co-produce and are reciprocally influenced by technological mediations. Designers create technologies with certain aims and use patterns in mind, whereby specific material features incite users to follow designers' intentions. Non-users also belong to this broad range of reflective actors, as are those who actively manifest their stance to *not* use a certain technology. Policy-makers are also concerned stakeholders who deliberate on the future use and regulation of a given technology (Verbeek, 2011).

The quote above summarizes the goal of the moral philosophy of technology, guided by the ideas of technological mediation. Namely, philosophy cannot prescribe the right course of action regarding a specific technology. Rather, it equips a broad range of stakeholders with the necessary means of inquiry—both empirical and philosophical—into specific human-technological practices, which empower them to make informed decisions about the design, use, non-use, and regulation of a technology under consideration.

I can now present a preliminary summary of the "moral" part of my conceptual exploration. I have argued that a relational approach to values best mirrors the background assumptions, scope, and goals of the technological mediation approach. Pragmatism allows the mediation approach to adopt a practice-based take on values, whereby they are enacted and manifested in lived human-technology practices. Adopting this perspective assumes a dynamic nature of values and their continuous developmental conception. Value dynamism denotes values as open to revision in view of the change in the sociomaterial environment that both enables and is oriented by them.

This leads to the next part of my theoretical exploration of moral mediation. Namely, what does the "mediation" part encompass regarding the idea of technologically mediated value dynamism? So far, the technological mediation approach has not explicitly dealt with the idea of value dynamics, instead focusing on how technologies co-shape situations of moral choice (Verbeek, 2011). The following chapter discusses how the mediation concept links technology and value dynamism by relying on its pragmatist origins. Apart from pragmatism, I also turn to other philosophical approaches that consider the relationship of values and technologies (e.g., Coeckelbergh, 2012; Vallor, 2016), particularly the approaches to technomoral change (Swierstra et al., 2009) and value change (van de Poel, 2021).

NOTES

1. As Ihde also notes, Husserl inherited this vocabulary from Descartes and all his work on reductions attempted to invert the distinctions implied in the early modern

"subject-object," "internal-external" vocabulary (2009, pp. 9–10). In view of this, Husserlian phenomenology, which Ihde deems as one of the defining for the larger field of phenomenology, propagated the use of this dualistic vocabulary, at least in the early stages of the field's development. Dewey's use of the environmental model of experiences and practices avoids this problem.

2. To be sure, Ihde (2009) identified what can be called "interrelational ontology" in both Dewey's pragmatism and classical phenomenology. Primarily in Husserl, Ihde suggests that interrelational ontology comes out in the notion of *intentionality*, consciousness of something, where "technologies occupy the "of" and not just be some object domain," mediating the meaning of consciousness itself (Ihde, 2009, p. 23). To reiterate, the interrelational ontology that Ihde identifies in Dewey rather concerns the environmental interrelational model of human experiences and practices with their material and social embedding. In this model, all counterparts are entangled, interdependent, and co-produce one another. The distinct feature of Dewey's interrelational ontology is that values, too, belong to his environmental model of experiences and practices.

Chapter 2

Technological Mediation of Morality

EXPLORING THE "MEDIATION" PART IN THE MORAL MEDIATION ACCOUNT

In the previous chapter, I suggested that the "moral" part in the moral mediation account assumes a relational and developmental account of values, sensitive to their sociomaterial embedding. Here, I explore how the "mediation" concept can incorporate such a dynamic and flexible definition of values while bridging it with technologies. To this end, I explore how the relational and developmental perspective on values is represented in various approaches to the ethics of technology, beyond technological mediation (e.g., Coeckelbergh, 2012; Swierstra & Rip, 2007; Swierstra, Stemerding & Boenink, 2009). I further rely on the pragmatism of Dewey, accompanied by the work of Walzer (1994), to clarify the different levels and stages of morality. Finally, I briefly correlate the technological mediation approach with two other approaches in the ethics of technology that also ground it in Deweyan pragmatism and explicitly acknowledge the co-evolving nature of technology and morality, namely the technomoral change approach of Swierstra (2011) and the value change lens of van de Poel (2021). This will allow me to delineate the scope of the technological mediation approach and its position on value dynamism.

As I have suggested above, exploring the ethical dimension of technologies is a well-established line of research in postphenomenology. Verbeek (2008, 2011) developed a moral mediation account within the technological mediation approach to demonstrate how technologies co-shape the moral perceptions and decisions of people. His ultrasound example showed how technologies mediate the moral decision-making of parents. This is especially significant in view of the ultrasound's ability to visualize potential genetic mutations such as Down syndrome. The specific manner in which the unborn appears on the screen enables different normative interpretations of it as a patient, a gendered individual, or a subject of care for prospective parents. This, in turn, presents different areas of moral interest and care for prospective parents, who, guided in part by what the ultrasound reveals, must make moral decisions about the future care of the child.

Highlighting the moral significance of technologies has become an important line in postphenomenological research. For instance, Wellner (2015) has demonstrated how cell phones mediate the moral attention of people, increasing the speed and scope of information processing and decreasing the threshold for focus, tolerance, and patience. Van den Eede (2015) has explored how self-tracking technologies mediate perceptions of the self and others, enabling dispersed human-technology intersubjectivity. Irwin (2018) has highlighted how the values of fairness, accountability, and transparency are reduced with the introduction of speech recognition algorithms. These examples demonstrate how postphenomenology accounts for the multiple ways in which technologies co-shape, challenge, or enable moral concerns.

To research how technologies mediate the value frameworks, however, the moral mediation account calls for an expansion. To date, postphenomenology has considered how technologies co-shape moral perceptions and choices. However, it has not highlighted how technologies can resurface a moral infrastructure for decision-making, reconfiguring the meaning of values and exposing their dynamism. The present conceptual exploration of technological mediation aims to do just that. Dewey's pragmatist account can help present the mediation approach as a specific bridge between technologies and values, thus making it sensitive to the idea of value dynamism. The two approaches are complementary, sharing fundamental theoretical assumptions regarding the relationality of experiences and practices. While the pragmatism lens emphasizes the environmental model of human practices, only in passing does it consider the role of its material setting. The mediation approach highlights the role of technologies in co-shaping human relations with the world but obscures the dynamic, back-and-forth nature of these relations. Through the pragmatism lens, I demonstrate how technologies can enable, re-affirm, challenge, or re-articulate values by highlighting the active role of each counterpart in dynamic human-technology-world relations. Before I further explore the contribution of pragmatism to expanding moral mediation, I inquire how other approaches in the ethics of technology have dealt with relational and developmental perspectives on values and what the mediation concept can infer from them.

ACCOMMODATING A RELATIONAL AND DYNAMIC VIEW ON VALUES

The phenomenon of value dynamism follows naturally from the relational nature of people and their sociomaterial environment. If value is understood as simultaneously enabling and being enabled by sociomaterial practices, a change in such practices can imply a different value manifestation. The moral mediation account, which embeds pragmatist considerations and acknowledges the mediating role of technologies, presents value dynamism as an inherent property of values in sociomaterial practices. Technological mediation is not alone in its pursuit of embracing moral hermeneutics by clarifying the interdependent developmental nature of values, as several authors work in similar directions (e.g., Coeckelbergh, 2012; Swierstra & Rip, 2007; Walzer, 1994; Swierstra, Stemerding & Boenink, 2009; van de Poel, 2021). Examining how other philosophers deal with the phenomenon of value dynamism can offer insight for the technological mediation approach.

For instance, Coeckelbergh (2012) emphasizes the necessity of understanding and exploring the evolving relations of people with the natural and material world and argues that only then can people form their relations consciously, guided from the inside, not by some external ephemeral force:

> We should shape our (new or already existing) relations with these [nonhuman] entities as these relations [. . .] are changing and growing. Instead of regulating what we do, instead of applying a Law or Code [. . .], we would do better to engage in the slow change of moral evolution and moral metamorphosis [. . .]. [T]here is no all-powerful and all-knowing Gardener—a god or we ourselves—which manages the moral order as a garden. There is change, but this change results from what we do in response to other entities and our environment, and what this environment and other entities do to us (2012, p. 204).

Coeckelbergh essentially appeals to taking seriously the relational nature of people with their sociomaterial environment because only their mutual co-shaping are identifying and living values possible. The author understands values as lived and experienced, not static or prescribed by some other authority, human or otherwise. Thus, when the relationality thesis is endorsed, the dynamic and relational nature of values follows.

Moreover, the ethics of relational growth that Coeckelbergh endorses oppose a passive analytic or observational stance toward values because such an external perspective on values would contradict their lived nature:

> We should not love wisdom in a Platonic way but act and find it in the world; there is no wisdom outside activity and experience. As Diogenes knew, we have

live wisdom and *live* value. Value is neither to be described nor to be created; it has to be lived (2012, p. 199, original emphasis).

In this sense, the technological mediation approach both takes the relationality thesis seriously and extends it a step further, attempting to operationalize it. Verbeek proposes to both *live* human-technology relations and, *in living them, to give them shape* by reflecting on how they develop (2011, pp. 156–158). The expanded moral mediation account, incorporating value dynamism, would help to reflect on how human-technology-world relations are both guided by certain values and reconfigure them.

Swierstra and Rip (2007) offer an interesting angle on value dynamism by suggesting that technologies can destabilize and provoke moral routines. New technologies can offer new possibilities that existing values and norms cannot offer satisfactory responses to. The ambivalence and uncertainty that new technologies represent can inspire reflection and an adjustment of normative views and moral intuitions. As Swierstra and Rip note, "Emerging technologies, and the accompanying promises and concerns, can rob moral routines of their self-evident invisibility and turn them into topics for discussion, deliberation, modification, reassertion" (2007, p. 6). New technologies often lead to moral disagreement, dilemmas, and the rearticulation of previously working norms. The way that technologies destabilize the moral landscape can allow us to inquire into the dynamics of morality and to determine how the (potential) presence of technologies mediates values.

I argue that framing technologies as moral provocateurs that spark moral reflection can explain one possible mechanism of value dynamism in the expanded moral mediation account. However, I would like to emphasize that focusing on the destabilizing role of technologies offers a limited view on the nature and scope of technological mediation because technologies can also enable new value practices and reaffirm existing value meanings. I developed this point more in detail later in this chapter in the discussion about mediation at the different levels and stages of morality. For now, it is important to further explore the view of technologies as moral disruptors, for it offers key insights about ethics and morality that the mediation approach can use.

The destabilization of moral routines, and the ensuing ethical discussions invite Swierstra and Rip to reflect on the nature of morality and ethics. In this, they build on the pragmatist scholarship of Dewey, who (1976b) suggested that morality is elusive and hidden because it is uncontroversial and accepted; it is essentially a compilation of values and norms that have proven to be effective and functional time and again, and as such, stabilized to the level of some abstraction that people need not review. For this reason, Dewey often refers to morality as "moral routines" and "moral habits," to highlight its generally accepted and tacit character.

According to Swierstra and Rip (2007), when technologies destabilize our moral routines, they thaw them from their (more or less) solidified form and enable us to reflect on them. Their consequent definitions of morality and ethics are as follows: "Whereas morality is characterized by unproblematic acceptance, ethics is marked by explicitness and controversy. Ethics is 'hot' morality; morality is 'cold' ethics" (2007, pp. 5–6). Ethics means reflection on morality: whenever one discusses moral conflicts or dilemmas or deliberates about what could be a good course of action, one *does* ethics. Or, as Swierstra and Rip frame it, "'ethical' is the articulation (including contestation) of what used to be morally self-evident" (2007, pp. 6). These ideas embed very similar principles as the technological mediation approach, perhaps not least because the authors were inspired by Dewey's moral philosophy. The above definitions of morality and ethics will guide my further explorations of the moral mediation account.

Adopting a relational and developmental perspective on values in the mediation approach produces several consequences. Firstly, it indicates the importance of reflection on human-technology-world relations that are both enabled by certain values and can reframe them. Secondly, morality is not a one-dimensional landscape but instead has different layers, or levels, which are typically dormant. For the present attempt to clarify the concept of mediation, this means that moral mediation does not concern values as single entities. Rather, values as building blocks of morality can comprise different dimensions that can manifest in response to the sociomaterial practice at hand. Finally, as Dewey elaborated, value dynamism closely relates to the degree of value stabilization, or the proven accumulated ability to be a working solution. I consider it important to clarify how technologies can mediate the different layers of morality and the different stages of its liquidity. I approach this subject by building on the ideas of Dewey (1976b), Walzer (1994), and Swierstra and colleagues (2009). This will consequently allow me to elaborate on the concept of mediation as sensitive to value dynamism and provide a starting point for where and how to study how technologies mediate values.

TECHNOLOGICAL MEDIATION AT DIFFERENT LEVELS AND STAGES OF MORALITY

Dewey specifically focused on the idea of moral malleability at different levels. Reviewing his ideas, with a view to the expanded moral mediation account, which can help to further elaborate the scope of mediation. In the rich scholarship of Dewey, the idea of moral malleability and value dynamism comes as a consequence of the relational co-shaping nature of people and their sociomaterial environment. To remind the reader, in this account, values appear not as moral finalities but as ends-in-view, open to the review of previously assumed moral avenues and the construction of new ones.

Another manner in which Dewey highlights moral malleability and dynamism comes to the fore in his revival of the Greek concept of potentiality. Dewey (1938, 1940) discusses potentiality as both a category of existence and an inseparable human quality of reinvention. Potentiality enables people to develop that which was previously latent and actualize their (moral) lives in novel ways through interactions with the world and its human and nonhuman entities (ibid., pp. 101—102). Dewey suggested that it is important to revive the Greek category of potentiality in modern moral philosophy to highlight the inherent ability of humans to reinvent themselves and their environment. In relation to values, potentiality represents the ability to review some stable notions that were used for years but cannot address present-day challenges. However, Dewey wanted to revive the classic notion of potentiality with a notable distinction by abandoning the idea of fixed moral ends in favor of moral ends also conceived as potentialities, actualizing only in relation to the environment:

> When the idea that development is due to some indwelling end which tends to control the series of changes passed through is abandoned, potentialities must be thought of in terms of consequences of interactions with other things. Hence potentialities cannot be known till *after* the interactions have occurred (Dewey, 1938/1940, p. 102, original emphasis).

Dewey famously rejected the idea of moral values as detached from the experiences of people, existing only as fixed external guidelines.[1] He similarly disagreed with accounts of values as being guided by the maximization of happiness and consequently getting lost in the cost-benefit calculus.[2] At the same time, Dewey acknowledged the value of overarching normative guidelines that consistently prove effective (e.g., 1922, 1930). To pursue reflective moral deliberation, one must occupy an established position and have some shared idea of where to begin. Dewey asserted that, the shared insights and overlapping conceptions of values are an objective fact of morality. However, contrary to deontologists (particularly Kant[3]) and utilitarians, Dewey suggested that shared conceptions of morality have experiential and temporal underpinnings and are the embodiment of multiple outcomes of concrete situations in the past that have proven right time and again. As such, their effectiveness in providing tentative guidance for future action has stabilized them. Hence, although Dewey admits to some form of overarching morality, he differs notably from traditional moral accounts. He first stresses the relative nature of overarching morality, which originates in problematic situations of the past, and secondly, he emphasizes morality's openness to change in the future should values prove unfit to direct further action (e.g., Dewey, 1922; 1929).

The work of Michael Walzer (1994) is critical to the moral mediation account because it further clarifies the relation between different levels of morality. Moreover, it suggests how morality can be both contextualized in experiences and practices and stabilized on an overarching level. Walzer suggests that two expressions of morality exist—minimal and maximal—that correspond to two levels of morality: thin and thick, respectively. According to Walzer:

> Philosophers most often describe [value dualism] in terms of a (thin) set of universal principles adapted (thickly) to these or those historical circumstances. [. . .] Morality is thick from the beginning, culturally integrated, fully resonant, and it reveals itself thinly only on special occasions (Walzer, 1994, p. 4).

Similar to Dewey, Walzer acknowledges and highlights the importance of thin morality and endorses its co-shaping relationship with the thick level. According to him,

> 'Minimalism' does not describe a morality that is substantively minor or emotionally shallow. The opposite is more likely true: this is morality close to the bone. [. . .] In moral discourse, thinness and intensity go together, whereas with thickness comes qualification, compromise, complexity, and disagreement (Walzer, 1994, p. 6).

Related to the ideas of moral mediation, Walzer's work suggests the importance of understanding the intricate relationship between the more abstract, or thin, and localized, or thick, levels of morality. They mutually inform each other, whereby "minimalist meanings are embedded in the maximal morality" (ibid., p. 3). This point echoes Dewey's critique of the detached individualistic conceptions of morality in Kant and his acknowledgment of the overarching morality as stemming from a mass of accumulated experiences. There is a lesson here for the expanded moral mediation account; namely, in the focus on localized human-technological practices, it is critical not to lose track of the larger sociocultural embedding, which is also normatively rich and telling.

In this regard, Walzer identified the importance of thin morality in attempting to thicken normative concerns. In an encounter with the unknown or the conflicting, the shared, thin morality allows people to collectively deliberate on the new phenomenon that caused (or can cause) a disruption. When people are confronted with something unknown or that conflicts with their conventional beliefs, they still recognize some features in it without knowing experientially, in elaborate, qualified terms, what it is. By this token, when technologies mediate values, they necessarily first confront their thin meaning, thickening it during the process of reflection and interpretation. This is also reflected in the exploratory case of Google Glass, briefly discussed in the introduction, whereby people had a shared, but almost elusive sense of privacy that in a process of collective inquiry and practical actions resulted in multiple understandings of it as they realized what was at stake with Glass.

Because people can unthinkingly rely on thin minimal morality, it is usually invisible and appears independent of the hurdles of human life. According to Walzer, only in times of crisis and disturbance do people discover that they have thin morality, and they consequently reflect and revise it in thick conceptions. Swierstra and Rip (2007) mirror Walzer's view when they frame technologies as moral disruptors, which marks a point of distinction between these approaches and the position of technological mediation. Namely, technology as a new unknown need not necessarily present a conflict or a crisis to reflect the moral hermeneutics of technological mediation. The moral mediation account can show how, in the course of value dynamism, moral frameworks are not only reaffirmed but also challenged. Augmenting the mediation approach with Dewey's emphasis on interrelational ontology in sociomaterial practices allows the examination of value dynamism *within* human-technology-world relations, and not only outside of them with crisis as an external destabilization. Moreover, it can also show how technology allows for new value meanings and practices. According to Dewey,

> Every object hit upon as the habit traverses its imaginary path has a direct effect upon existing activities. It reinforces, inhibits, redirects habits already working or stirs up others which had not previously actively entered in. In thought as well as in overt action, the objects experienced in following out a course of action attract, repel, satisfy, annoy, promote and retard. Thus deliberation proceeds (Dewey, 1922, p. 192).

Thus, Dewey suggests that an encounter with a new phenomenon can enable new and review existing moral habits, both projectively and in practice. From both the mediation perspective and Deweyan pragmatism, it follows a crisis that is not a precondition to value dynamism. Contrary to Walzer (1994) and Swierstra and Rip (2007), the technological mediation approach, accompanied by pragmatism, maintains that value dynamism can also manifest in non-crisis situations, with technologies framed not only as disruptors but also as enablers of moral practices.

Nevertheless, understanding the co-shaping relationship between the thin and thick levels of morality is relevant to the discussion of moral mediation. It suggests how destabilization, crisis, or conflict induced by technologies as moral disruptors can offer one method to detect value malleability. However, it is critical to remember that a reference to crisis might not always be necessary to illuminate the moral hermeneutics of technologies. I explore other ways to identify and reflect on value dynamism in the chapter that follows, where I study how people accommodate technologies in their referential frameworks (see chapter 3).

What the discussions above clarify is the importance of considering different levels and stages of morality when examining value dynamism. However, the difficulties that emerge when explicitly considering the role of technologies in value dynamism also become visible. Tsjalling Swierstra's approach to technomoral change was the first to highlight the active role of technologies in the constitution and change of values (2011), while the approach of Ibo van de Poel (2021) was the first to demonstrate and explicate specific mechanisms through which technologically induced value change unravels. To delineate the ambitions and scope of the mediation concept in moral hermeneutics, in the next section, I wish to briefly correlate the technological mediation approaches with these two approaches.

EXPLICITLY CONSIDERING TECHNOLOGIES IN MORAL HERMENEUTICS: THE TECHNOLOGICAL MEDIATION APPROACH VIS-À-VIS TECHNOMORAL CHANGE AND VALUE CHANGE APPROACHES

The philosophical approach to *technomoral change* suggests that the normative frameworks of people are not static but co-evolve with the introduction of new technologies (Swierstra et al., 2009). It draws inspiration from the pragmatist ethics of Dewey and considers the specific role of technology in value change. Technology introduces new courses of action and, with this, opens new moral avenues or invites a review of the old ones (Swierstra, 2013; Vallor, 2016). At the same time, providing new courses of action inevitably means relying less on the old ones and, correspondingly, relying less on the values they embody. The accents we place on values and how we interpret them also shift. For instance, good manners in the 1960s dictated offering your visitor a cigarette. Today, this would be considered inappropriate since smoking does not correspond with a healthy lifestyle value (Swierstra, 2011). In summary, the technomoral change approach is interested in how value changes occur in relation to technologies.

To explore how values evolve over time and with the introduction of technologies, Swierstra and colleagues (2009) rely on the robustness and clarification of a given value as an entry point. According to the authors, the robustness or stabilization of values passes across three levels: the macro-level, meso-level, and micro-level. The macro-level reflects only gradual developments of fundamental and abstract normative principles (such as flourishing), which do not easily change. Their change, even when induced by technologies, is not quick because they have proven suitable guidelines for ethical problems in different contexts over time. The meso-level of technomoral change specifies a value to some extent by reviewing it in specific practices (for instance, the well-being of a child). Because ethical problems are subject to cultural and social interpretation, different interpretations may exist on how to enact the value in question. Thus, closure regarding what it means never really occurs. Finally, on the micro-level, the practical requirements of a situation require concrete ethical questions and decisions. For example, a concrete ethical question concerning the well-being of a child could be the following: Should the prospective parents perform a full genome sequencing on their newborn baby? Because practical requirements and options often change, fueled by cultural traditions and individual moral intuitions, the normative micro-level is not likely to stabilize. Thus, the technomoral change approach uses the pyramid of value robustness as an indicator for how some values are more prone to change than others. It also suggests that the micro-level of concrete ethical questions is the most favorable for exploring the dynamics of technomoral change.

Although the technological mediation approach and the technomoral change approach similarly endorse the co-shaping nature of values and technologies, they do so for different purposes. The technomoral change approach was originally developed to deepen and substantiate policy-making discussions regarding the future of a given technology in society (e.g., Swierstra et al., 2009; Boenink et al., 2010; Swierstra & Rip, 2007). As such, it adopts a broader societal lens to explore how technologies change values. The technological mediation approach focuses more on the individual level to inform the practices of technological use (Verbeek, 2005) and, to a lesser extent—design (Dorrestijn, 2012). The mediation approach builds upon post-phenomenology with its dedication to lived experiences and a first-person perspective. As such, it explores the moral mediation of new technologies, beginning with individuals: how people appropriate new technologies and make them meaningful, and how technologies mediate the concrete experiences and practices of people. The technomoral change approach does not at all exclude an individual viewpoint. However, in an attempt to provoke group ethical deliberations, it must scale up from the individual level to present broader generalized concerns, to which many individuals can relate. At the same time, the mediation approach, by developing a new focus on value dynamism, can potentially extend beyond individual concerns and become useful for discussions at a larger scale.

Another point of distinction between the approaches could involve the object of interest. Whereas the technomoral change approach explicitly focuses on a change in values over time, the mediation approach explores how technologies mediate values and induce their dynamics in the present. As the pyramid of value robustness indicates, the technomoral change approach considers change in values over a broad temporal trajectory. The mediation approach focuses on lived practices to show how different dimensions of values materialize in the present. From this temporal perspective, the technomoral change approach's scope is arguably larger than that of mediation, which scrutinizes technologically mediated value dynamics as they occur during human encounters with technologies. At the same time, the somewhat narrower focus of the mediation approach allows the exposure and scrutiny of the dynamics of value change itself, which remains underexposed in the technomoral change approach.

These distinctions between mediation and technomoral change are not clear-cut, and both approaches could potentially venture into thinner or thicker, or present or future-oriented, domains. Kiran, Oudshoorn, and Verbeek (2015) have suggested that the technological mediation approach can also contribute to policy-making on new technologies and anticipate future moral mediations, given that technologies help co-shape the ethical debate around themselves. Similarly, Swierstra (2016) has clarified that the technomoral change approach requires an elaboration of daily lived messy morality and that policy-level deliberations require such thick substantiations. Both approaches mutually inform each other since they represent different aspects of the same process. No technomoral change would exist without technologically mediated value dynamism, while at the same time, value dynamism, however foundational, is a first step in the larger process of value change.

Recent explicit attention in the field of ethics of technology to the role of technologies in morality resulted in another noteworthy account, that of the technologically induced *value change* (van de Poel, 2021; van de Poel & Kudina, 2022). Originating as a critique of the Value Sensitive Design account (Friedman et al., 2002), van de Poel suggested that the idea of designing technologies with specific values in mind is shortsighted, albeit noble. More specifically, it disregards the fact that values are not pre-given static entities and rather evolve with their sociomaterial environment, changing and giving place to a new value conceptualization (2021). With this, van de Poel set out to sketch the foundation of value change by explaining what makes it possible and suggesting the mechanisms of value change, examining different pathways along which value change can unravel.

Notably, the account of value change shares with the mediation account and the technomoral change approach its ontological grounding in Deweyan pragmatism. Beyond sharing the idea of values being lived and relational, co-evolving with their social and cultural environment, there are some notable distinctions. According to the value change account, value change, when it is technologically induced, is a result of morally indeterminate situations that technologies in our midst confront us with. This triggers an inquiry into what is morally at stake (i.e., reconfiguring an indeterminate situation of general discomfort and moral puzzlement into a concrete problematic one), examining whether the existing value conceptualizations help us to practically resolve the current indeterminate situation or whether there is perhaps a need to review the existing moral pointers or even give way to the new ones.

This account also gives a more nuanced definition of values, grounded in the pragmatist tradition. Firstly, values help to evaluate the situation at hand, and in that sense, they may be understood as devices for evaluation. Secondly, they are hermeneutic in nature, resulting from prior human experiences with other problematic situations, and are thus open to review when they no longer can resolve the current problem. Lastly, their hermeneutic nature grants values a degree of shared moral background in society, providing a helpful overlap when starting a new moral inquiry and giving a degree of justification. Distinguishing such features and functions of values allows van de Poel to define them as evaluative devices that are based on the prior experiences of people and that help them to identify, examine (and sometimes, resolve) the morally problematic situations, oftentimes induced by technologies (van de Poel & Kudina, 2022).

In an important distinction from the technological mediation account and the technomoral change approach, van de Poel's approach examines and proposes the specific ways in which value change can proceed, namely value dynamism, value adaption, and value emergence (Ibid.). Value dynamism occurs when there is an interpretation of the existing value frameworks as a result of an inquiry into a morally problematic situation. Thus, the existing normative frameworks still hold in general, but they have to be re-interpreted in light of some new factors from the sociomaterial environment to be able to properly address them and give some clues for action. For example, the introduction of new digital technologies, such as the video doorbell ring by Amazon, is often accompanied by privacy concerns. While the companies behind technologies often emphasize the individualistic notion of privacy as control of one's information, the new practical situations confront the users of technologies, such as Ring, with alternative privacy dimensions. For instance, the use of Ring also concerns the privacy of the passerby and those in public locations that happen to be within the constant video (recording) reach of the Ring doorbell. This renders the control of information conception of privacy ineffective, at least regarding the social dimension of privacy, causing a morally problematic situation and calling for a reinterpretation of privacy to be able to address it. As a result, some owners of Ring opt for its timed selective use, review the device settings to make it more privacy-sensitive toward others, or choose not to use it for some time. These reviewed uses of the device do not render privacy obsolete, but quite the opposite; they demonstrate the active and reflective ways in which people try to deal with its multiple dimensions and new interpretations. This is an example of value dynamism because we have a value that is carried over from previous situations and is reworked in a new morally problematic situation, induced by technology, that renders the previous conceptualization ineffective.

When such value reinterpretations remain localized or specific to a given technology, we can refer to them as individual value dynamics. However, when they escalate to a general societal debate and trigger large-scale institutional adjustments concerning other types of technologies, we may speak of a broader process of value adaptation (Van de Poel & Kudina, 2022). The more collective level of change during the process of value adaptation also refers to a shift in priorities or concerns that underpin a certain value, prompted by a morally salient situation. For instance, the introduction of digital technologies in the 1980s and 1990s was closely correlated with people's valuing of unrestricted memory and the desired ability to access any piece of information from any point in the world on demand. However, the detailed use practices with digital technologies and the Internet revealed many new indeterminate and problematic situations due to the perpetual digital footprint people leave behind whenever they use online services and the difficulty to remove such digital traces. This resulted in a multitude of individual information removal requests from the search engines and prominent court cases trying to resolve these situations (Mayer-Schönberger, 2009). In a span of several decades, the Western European context underwent an important moral shift from perpetual remembering to the value of forgetting and being forgotten by the Internet, mirrored in the newly manifested legal right at the EU level, "A right to be forgotten" (Jones, 2018). This case represents a mechanism of value adaptation because we witness a collective shift initially in individual priorities regarding the situation of information management. While the values of remembering and forgetting have always accompanied such practices, here we observe a clear adjustment in priorities away from the value of remembering because it could not properly address the changing sociomaterial practices. This does not mean that such a value adjustment resolves all the morally indeterminate situations regarding digital practices, but it suggests that the value of forgetting, at least for the time being, satisfies the public as a moral orienteer, until a new collective inquiry starts.

Finally, the third mechanism that the value change account proposes is value emergence (Ibid.). It underscores a situation when the result of working on a problematic situation is not satisfied by a reworking of the existing values and requires a distinctly new one to satisfy the existing demands. When this occurs, the new value undergoes a transformation from a response to a specific one-time situation to a gradually recognized one, prompted by similar problematic situations time and again. A case in point here could be the value of sustainability that gradually entered global attention and priority from originally small-scale localized situations. In the span of human history, localized environmental disasters, often due to human-technological interventions (e.g., deforestation and coal mining), produced sustainability as a proto-value, an ad hoc appeal that only very gradually turned into a general, indeterminate situation at a global level. An unprecedented rate of technological progress in the twentieth century prompted an escalation of the value of sustainability to a global arena, underpinned by the catastrophic experiences of world wars and nuclear accidents. The global penetration of mass media further contributed to the generation of public awareness and knowledge of the environmental state of affairs, demonstrating not just the local but the global scale of similar ecological concerns. We could refer to this situation as a case of value emergence because there is a clear evolution of it from an old localized concern to a consolidated global concern, particularly in the twentieth century. It was also not a first response value to crafting a harmonized relationship of humanity with its environment, but a gradual processing of existing values that could not in themselves help resolve the situation. Over time, the localized call on the value of sustainability became increasingly recognized internationally, both as a moral guideline and a desired outcome.

In sum, van de Poel's account helps to clarify the phenomenon of technologically induced value change by identifying morally indeterminate situations as the root cause of value change and specifying the ways in which technologies can help to provoke it. In this, it offers a clearer and more precise account of value change vis-à-vis the technomoral change approach of Swierstra (2013). Notably, the two accounts originated for different purposes, which might explain the degree of nuance they offer: the former—to guide responsible design practices, and hence requiring more detail; the latter—to help institutional deliberation regarding new technologies, and hence not offering one-size-fits-all mechanisms and presenting a set of deliberative tools instead.

Additionally to these two accounts, the technological mediation approach seems to consider moral mediation as an instantiation of moral experiences, as a way to inform people of what might be at stake in their practices with technologies. Its phenomenological focus is more geared toward clarifying how technologies might change our moral perceptions and co-shape the avenues for moral actions than detailing the infrastructure of these changes. In this case, there is a strong overlap with the value change account, whereby value dynamism appears in both approaches to clarify the experiential step of morality-in-the-making as induced by technological factors. However, while in the technological mediation approach this form of moral mediation of technologies serves to clarify the hermeneutic nature of morality and its engagement with the human-technology-world environment, in the value change account it represents only the first step in their moral co-evolution, followed by value adaption and emergence.

All three accounts of technomoral change, value change, and technological mediation thus definitely share the focus on the moral change due to technologies, although they differ as to the starting point of their exploration. While the value change account seems to be more interested in the mechanics and principles of moral hermeneutics, the technomoral change account is more prone to exploring its visible manifestations, i.e., the soft impacts resulting from value change, and the technological mediation approach is primarily concerned with moral hermeneutics from the angle of change to human practices and experiences with technologies.

CONCLUSION

For now, what the discussion in this chapter clarifies is that morality and ethics are two sides of the same coin. Morality presents the accumulation of workable normative means and guidelines to direct present action, and ethics lends itself to a reflection on these tacit moral rules (Swierstra & Rip, 2007). To transform morality into ethics requires some sort of moral disruption. Technologies can convert dormant moral views into reflective ones by confronting them with new situations and challenging them in other ways (Ibid.). However, technologies not only disrupt moral routines but also foster new moral understandings and co-shape ethical debates within human-technology-world relations (Verbeek, 2011).

I have also endorsed the view that morality is multilayered, consisting of thin and thick levels, conceived as abstract normative guidelines and as experientially rich normative contestations, respectively (Walzer, 1994). Abstract moral views can act as entry points to thicker elaboration. The lesson for the moral mediation account here is that to explore value dynamism, both levels of morality are important.

The pragmatist lens of Dewey has been particularly helpful in developing the conceptual vocabulary for the technological mediation approach to account for the phenomenon of moral hermeneutics. To be able to more deeply consider value dynamism and change, I have argued for the need to return postphenomenology to its pragmatist origins. As it currently stands, the mediation approach primarily focuses on the micro-practices of people with technologies. The pragmatist approach provides this with more depth, suggesting that in those practices, the sociomaterial world is enacted, including the value frameworks. Returning to pragmatism allows the expansion of the postphenomenological approach to the idea of moral hermeneutics in general and value change in particular, offering an expanded version of the technological mediation approach as presented in *Moralizing Technology* (Verbeek, 2011). At the same time, it also suggests where and how value dynamics can be observed. Namely, the interrelated nature of values and sociomaterial practices hints at the profound hermeneutic dimension of values that must be clarified, the one that transcends situations of crisis. As such, enriching postphenomenology with Deweyan pragmatism allows the substantiation of its interrelation claims at an ontological level, supported by an environmental model of sociomaterial practices, and the production of a relational dynamic account of values as both enabling and being co-shaped by human-technology-world relations. Using Dewey's pragmatist lens in this manner facilitates the first steps toward understanding postphenomenology as a distinctly moral philosophy of technology.

The conceptualization of values as relational and the focus of the mediation approach on lived experiences invite the location of values in human practices with technologies. However, this is not an easy task. As explained by Mol (2002) and Dussauge and colleagues (2015), there is no direct access to values in practices, as there is no mental beeline to locating them in the minds of people. This requires a process of interpretation related to existing human experience and the larger sociocultural embedding. This is precisely what the line of appropriation in the moral hermeneutics account aims to clarify, as followed in chapter 3. There, I will explore the hermeneutic dimension of values in sociomaterial practices through the process of appropriation. The following chapter elaborates on how examining the dynamics of appropriation can deepen our understanding of moral hermeneutics while explicitly linking it to technology.

NOTES

1. To reiterate: Dewey wants to return to the Greeks in building his empirically-grounded moral philosophy because they endorsed the critical reflective choice and the relational nature of all forms of life, but he wants to change its assumptions regarding the fixed (moral) ends in favor of the innovation and discovery of human good. Dewey drew inspiration from Plato's cosmology, but opposed its fixed and finite order. Plato suggested that for harmonious interaction of people with the whole of cosmos to occur, there is but one social order and one structure of human nature, and any alteration from this predetermined structure would yield chaos and disaster. According to Dewey:

> The return must abandon the notion of a predetermined limited number of ends inherently arranged in an order of increasing comprehensiveness and finality [. . . recognizing] that natural *termini are as infinitely numerous and varied as are the individual systems of action they delimit*; and that [. . .] new individuals with novel ends emerge in irregular procession. It must recognize that limits, closures, *ends are experimentally or dynamically*

determined, [. . .] that they [moral ends] intersect everywhere; that it is uncertainty and indeterminateness that create the needs for and the sense of order and security (Dewey, 1929, pp. 395–396).

2. Dewey opposed the idea of reflective deliberation being reduced to the utilitarian calculus of attaining maximum happiness:

> Deliberation is irrational in the degree in which an end is so fixed, a passion or interest so absorbing, that *the foresight of consequences is warped to include only what furthers execution of its predetermined bias*. [. . .] The office of deliberation is not to supply an inducement to act by figuring out where the most advantage is to be procured. It is to resolve entanglements in existing activity, restore continuity, recover harmony, utilize loose impulse and redirect habit (Dewey, 1922, pp. 198–199).

3. Dewey's main issue with Kant consisted in Kant's individualistic conception of moral philosophy, originating in the pure reason of single individuals and disregarding the fact of moral disagreement (1922). Hence, moral rights and ought's can be understood rationally by all human beings, and when understood rationally, no disagreement between such conceptions can occur. Hence, there is no need to consult actual human experiences. Dewey took issue with both experientially unaided individual reason and the supposed lack of moral disagreement. Specifically, he disagreed with how they presupposed a conception of nature as rational and as such, perfectly accessible to and revealed identical to human reason. Dewey's analysis of nature in its diverse socio-material manifestations, human beings, and moral ends as ends-in-view, all entangled and interdependent, suggests that moral conflict and disagreement are inherent and persistent in the relational order of things. Therefore, Dewey opposed Kant's moral judgments as absolute and stemming from single individuals without regard for the context and other individuals.

Chapter 3

Technological Appropriation and Moral Hermeneutics

DEFINING APPROPRIATION

To remind the reader, the larger goal of this book is to understand how technologies mediate human values, or to attempt the study of moral hermeneutics. I am curious as to how moral concerns and perceptions manifest in relation to technologies and how existing normative ideas undergo rearticulation, facilitating moral decisions. Chapter 2 allowed us to position values as relational to the sociomaterial practices and as dynamic. This chapter further explores the connection between values and technologies by scrutinizing the idea of appropriation, originally suggested by Verbeek (2015, pp. 101–103), as potentially linking morality and human sensemaking regarding technologies. I would like to scrutinize that intuition and see how it can be useful for understanding the moral hermeneutics. More specifically, I want to get a better understanding of appropriation by zooming in on both its projective and practical dimensions, when the technologies in question already exist on the market, and projective, or more cognitive, appropriation when technologies still exist primarily as visions and promises, promotional campaigns, ethical concerns, and opportunities. Taking cue from Grunwald (2016), in order to understand what may be morally at stake with a certain emerging technology, it is important to understand the visions of the world it promotes, the narratives it evokes, and the promises and dangers materialize in this regard—even when hardly anyone has actual experience with the technology in question. I will refer to this collective phenomenon of positioning technologies into the interpretative frameworks of people as "human appropriation of technologies." I will also show how exploring the way people appropriate technologies (projectively or in practice) can be helpful for understanding the dynamics of moral hermeneutics, showcasing morality-in-the-making while explicitly linking it to technologies. Effectively asking the question: "How is making technologies our own related to making moral sense?", this chapter explicitly links the study of morality-in-the-making to the philosophical field of hermeneutics. The first steps in this endeavor require investigating the scope, theoretical standing, and significance of the concept of appropriation, which I will do next.

The Lens of Domestication Studies

The concept of appropriation is not new, and it has a firm standing in the field of Science and Technology Studies (STS), particularly in the theoretical and practical framework of domestication (e.g., Silverstone & Hirsch, 1994; Sorensen, 2006; Berker, Hartmann, Punie & Ward, 2006a). Domestication refers to the process of "taming" new technologies and adopting them into households and other sociocultural units. Appropriation appears here as the first stage of the process of domestication, concerned with an initial attribution of meaning to a new technology before it is (or is not) fully adopted and integrated into the household. Thus, to clarify the meaning and ambition of appropriation as originating in the technological mediation approach, it is first of all important to distinguish it from the meaning and scope of this concept from the perspective of domestication studies.

Domestication as a concept, method, and theory emerged in the late 1980s, suggesting that the meaning of a certain technology is negotiated by both the inside (i.e., the domestic, the household) and the outside (i.e., the market, objective economy), making them mutually constitutive (Silverstone & Hirsch, 1994). According to Sorensen (2006), "it presupposed that users played an active and decisive role in the construction of patterns of use and meanings in relation to technologies" (p. 46). As such, domestication is a relational activity where both the user, her environment, and a new technology must become accustomed to one another; renegotiate certain boundaries, routines, and meanings; and conjure new ones. Carter, Green, and Thorogood (2013) have referred to domestication as an interactive process that "involves symbolic work (the creation of meaning), cognitive work (learning) and practical work (for example, changing patterns of use and daily routines)" (pp. 348–349). As a part of the Social Construction Theory in STS, domestication studies focus on the negotiation of social rules and routines as well as on the power and control that users engage in to adopt a new technology (e.g., Berker, Hartmann, Punie & Ward, 2006a).

To perform a domestication analysis, it requires considering a new technology in context, studying the daily routines, and practices of its consumers and understanding their larger sociocultural embeddedness. To achieve this, domestication studies have proposed approaching the process in four stages—appropriation, objectification, incorporation, and conversion. A qualitative empirical lens would accompany them to "encapsulate the nuances of consumption and the way that users inscribe artefacts with meaning to give them a place in a network of the home and everyday life" (Berker, Hartmann, Punie & Ward, 2006b, p. 6).

Originally, appropriation referred to the context of consumption and assimilation within the moral economy of the household:

> An object—a technology, a message—is appropriated at the point at which it is sold, at the point at which it leaves the world of the commodity and the generalized system of equivalence and exchange, and is taken possession of by an individual or household and owned. It is through their appropriation that artefacts become authentic (commodities become objects) and achieve significance (Silverstone, Hirsch & Morley, 1994, pp. 18–19).

Appropriation, thus, denotes the process of encountering a generic object and making it one's own, revealing how a generic object from the outside realizes its symbolic and physical potentiality in the rich sociocultural world of an individual or a household. Silverstone, Hirsch, and Morley further remark that "Appropriation reveals itself in possession and ownership" (1994, p. 19), thereby emphasizing the practical use and physical presence of a product for appropriation to occur. Or, in the words of Carter, Green, and Thorogood, "Appropriation at its simplest occurs when a device is purchased and users take it home" (2013, p. 350).

Such a narrow, strict meaning of appropriation has blurred as domestication studies evolved, ultimately appearing to reach the scope and level of generalization of the process of domestication itself: "Domestication involves the appropriation of the new into the familiar, or [. . .] as a process in which that appropriation is attempted" (Silverstone, 2006, p. 245). However, the meaning of appropriation as the first step in the domestication process has prevailed to date in many case studies that perform domestication analysis (e.g., Lim, 2008; Richardson, 2009; Carter, Green & Thorogood, 2013; Bertel, 2018).

In summary, within domestication studies, the concept of appropriation denotes the process of acquiring a new technology, bringing it into the household, and attributing it with a preliminary meaning. Making sense of a technology here belongs to the level of collective entanglement rather than single individuals (with an exception of Sorensen [2006], who also focuses on individual domestication), because multiple sociocultural contingencies, crucial for overall domestication, best manifest at the level of a family household, workplace, or society. Appropriation analysis, just like the overall domestication study, is descriptive by nature and has a sociological standing. However, it provides a rich detailed grounding for normative conclusions for those who seek them. Finally, the appropriation stage in the domestication study concerns the overall consumption and commodification process, impossible without the actual presence of a technology in the marketplace and, consequently, in the household. Clarifying the origins of the term through the domestication lens allows me to now outline how appropriation can be defined with the help of the technological mediation approach and in view of the ambitions of the moral hermeneutics study.

The Lens of Technological Mediation

In the approach of technological mediation, the concept of appropriation acquires a postphenomenological grounding and the underlying ideas of technological mediation. Phenomenologically speaking, appropriation is a hermeneutic activity. During appropriation, people interpret a technological phenomenon and integrate it into their existing frameworks of understanding, necessarily updating them. The process of appropriation is not confined to the theoretical domain but instead resembles the mutually informing symbolic, cognitive, and practical activities that underlie domestication (Carter, Green & Thorogood, 2013). Therefore, in the technological mediation sense, appropriation refers to both the projective and practical activity of sense-making, implicitly or explicitly attributing meaning to a new technology. The foregrounding and revision of normative ideas and intuitions, moral routines, and habits occur in this multidimensional process.

Comparing the ambitions and scope of the appropriation concept from the points of view of domestication studies and the technological mediation approach allows us to outline some resemblances and considerable differences. Appropriation denotes a hermeneutic dimension of mediation, or how people take up and attribute meaning to technological mediations. It is an intentional process, always directed at a specific artifact within embodied sociocultural experiences. The appropriation concept in domestication studies predominantly analyzes the post-factum practical adoption of technologies in the daily lives of people, particularly in the setting of their homes. The appropriation concept in the technological mediation approach, however, concerns the projective dimension of this process as strongly as the practical one. Projective appropriation explains how people, confronted with an uncertain and ambiguous technology, make sense of it and attribute it with meaning by relying on their past experiences, sociocultural embedding, and information from various sources. Projective appropriation demonstrates that people can experience a certain technology long before practical exposure. Phenomenologically speaking, projective appropriation represents the hermeneutic circle in action (Gadamer, 1975/2004), whereby people comprehend an unknown by projecting their own histories and personalities onto the sociotechnical environment, continuously revising preliminary meanings with new information and practical experiences. The appropriation concept in the technological mediation approach, in short, denotes how people, both explicitly and implicitly, develop a relation with technology.

Another difference concerns the target group of appropriation. In the domestication study, the primary focus is on the group level: the level of the household or society. In contrast, the phenomenological origins of the technological mediation approach incline it to favor a micro-perspective, which considers the rich lived experiences of single individuals. Therefore, large-group generalizations are not the primary goal in the appropriation study under the technological mediation approach (although they can be); this cedes the focus to informed perspectives of how the construction of meaning occurs in in-depth individual cases.

Similarly, because projective processes occurring during appropriation in the technological mediation sense (e.g., conceptualizing, comparing, fostering new meanings, reconfiguring existing ones, etc.) are equally important as the practical ones (e.g., physically approaching a technology in question, understanding it through use, etc.), the physical presence of a technology is not as essential for appropriation here as it is for domestication studies. Although the experience of using a technology provides a richer, more balanced, and more nuanced canvas against which appropriation occurs, the possibility of using the technology is not always there: for instance, when the technology in question is released to a limited number of users (consider the Explorer program for Google Glass) or is still on the brink of introduction (e.g., the sex selection chip). In the current age of technological innovation, new technologies appear daily, whether in a physical or digital form or in the shape of technological visions. In this context, people must often deal not with the technology itself but with the promises, hopes, visualizations, video presentations, scenarios, debates, concerns, and fears surrounding it (e.g., the threat or the promise of artificial intelligence). Before the technology actually enters the market and the household, people already possess an idea regarding what it is and how it fits (or does not fit) with their mindset, life goals, habits, and moral landscape. In short, people have already appropriated it projectively. The absence of rich experiences with technology and its emergent status should be no excuse to disregard the visions and narratives it invokes. On the contrary, echoing Grunwald (2019), to facilitate its responsible introduction in society, it is imperative to examine the meanings people attribute to a certain technology based on the hopes and fears it evokes. Engaging with these narratives will provide rich ground for, among other things, examining potential value conflicts and opportunities, as well as resurfacing what morally guides people in their anticipated practices with the emergent technology in question. This is exactly what the projective dimension of technological appropriation attempts to do. Thus, contrary to the domestication studies, the actual presence and use of a certain technology is not a prerequisite to its appropriation in the technological mediation sense. Rather, the study of technological visions would be a first step in understanding how people accommodate (emerging) technologies.

Finally, a crucial difference between the two takes on appropriation is an explicit focus on its normative dimension from the technological mediation perspective. While domestication studies do encounter the changing moral routines of people with the adoption of new technologies (e.g., Berker, Hartmann, Punie & Ward, 2006a), this is more of an incidental normative finding of the otherwise sociological study, concerned with the overall process of fitting a technology in the daily lives of people. In the technological mediation approach, the concept of appropriation is intimately linked to moral hermeneutics, or the study of morality-in-the-making.

The approach of technological mediation establishes *that* the norms and values of people are not divorced from the technologies around us. Consider Verbeek's (2008) analysis of how ultrasound technology co-shapes prospective parents and their yet-unborn children in different (normative) contexts, providing certain responsibilities and roles. While this example provides a convincing illustration of technology's mediating role in morality, it does not explain *how* such a mediation occurs. Consider another example, adapted from Swierstra (2011). In the 1950s, when one hosted guests, social norms dictated that guests be offered a cigarette as a sign of care and politeness. Today, however, offering a cigarette could be considered not only impolite but harmful to the health of guests. How is it that the same technology can foster radically different norms regarding its use? Furthermore, what is it that ensures such value dynamism? I hope to answer these questions with the help of the appropriation concept.

In the cigarette example, it seems that normative ideas about cigarettes have changed with additional knowledge that people have gained from their use over time, while the cigarettes themselves have remained practically the same. Appropriation denotes a process that captures the dynamic balance between the existing experience, perceptions, and knowledge that people possess, including normative views, on the one hand, and the cumulative unknown that a new technology represents in a given context, on the other. I suggest that an attempt to understand and interpret a technology is inherently connected to the shaping and negotiation of the norms and values of people. The detailed dealings of such interpretative processes of appropriation deserve further inquiry.

A conceptual clarification offered so far allows for a preliminary understanding of the appropriation concept by suggesting that it represents a sense-making activity that involves the interaction of (at least) three actors: people, with their existing knowledge and beliefs; technologies, representing a phenomenon that requires the attribution of meaning and its integration into the existing frameworks of understanding; and the sociocultural world, as an active context against which the human-technology encounter occurs.

At least two important conclusions follow from this preliminary definition. First, appropriation is always an intentional activity directed at a specific technology. Also, as the cigarette example suggests, appropriation proceeds both projectively and practically to constitute a single mode of appropriation. Based on projective appropriation, one may choose to review the practical use or refrain from using the technology in question altogether. In this broad sense, technological appropriation never fails. This leads to the second conclusion, namely, that the three dynamic and interrelated elements of the appropriation process prevent it from being a static, once-and-for-all event. Once a change has occurred in one of the elements constitutive to the appropriation process that does not fit the interim appropriation mode, a new or revised meaning is produced to better accommodate the situation. The stability of the appropriated technology, or the preliminary meaning bestowed upon it, depends on the interaction of people and the technology in question in a specific sociocultural setting. However, this remains a dynamic and fluid process. I suggest giving these conceptual elaborations a reality check by applying them to the case study of Google Glass, a socially contested augmented vision technology originally introduced at large in 2013 and reintroduced as a niche application around 2019. Analyzing how people appropriated Google Glass can serve to clarify some of the conceptual assumptions above and further flesh out the research direction for the study of moral hermeneutics.

THE NEED FOR A PROOF OF PRINCIPLE

Before plunging into deeper theoretical investigations and thorough empirical case studies of the concept of technological appropriation and its relation to moral hermeneutics, I would like to have a proof of principle of whether a connection between appropriation and moral hermeneutics stands. Aided by an exploration of the use of Google Glass online, I suggest that the mediation approach can do more than it has to date in the ethics of technology, beyond demonstrating that technologies mediate morality by co-shaping the situations of moral choice. I think it is also possible to determine how the mediation approach reveals the dynamism in the meaning of values themselves and establish the principles of moral hermeneutics that underlie this. Such an expanded moral mediation account would not only involve technologies having ethical implications and steering ethical behavior but would also be about technologies mediating the value frameworks. A tentative empirical exploration will allow me to verify whether this occurs through the study of technological appropriation.

At this preliminary test stage, I refer to Google Glass as an example because in 2015, after its introduction, it provoked much discussion, both in the media and in everyday life, regarding the value and meaning of privacy. As of 2022, it inspired the creation of similar augmented reality goggles that currently gain popularity across the world (e.g., the smart glasses Spectales by Snap or Bose Frames). One of the main reasons why the first working version of Glass provoked adamant public discontent is that it had no way to notify surrounding people that its user was taking pictures or recording video. In Europe, in the context of the still-emerging General Data Protection Regulation, privacy in the case of Google Glass was framed predominantly with the goal of protecting the data of citizens. This almost automatically centers all privacy discussions around the control of information. However, informed by users of various digital technologies, online discussions, and own experiences, I knew that privacy comes in many flavors.

An intuitive case in point of technological appropriation and morality for me was the example of stickers over laptop webcams and microphones. Often, it is a self-imposed creative solution in the name of privacy to prevent the malicious interception of video and audio channels. I sensed that such a creative appropriation of technology could reveal much about the meaning of privacy both in general and in relation to specific technologies, as well as about the roles and responsibilities of their users. More specifically, understanding how people think about their gadgets, foresee engagement with them (projective appropriation), and go about using them (practical appropriation) can demonstrate what people value. Moreover, such an understanding can show how a technology can confront people's values, forcing them to find creative roundabouts to preserve that which is valuable, give it up, or forge an alternative. If tiny cameras on everyday devices such as laptops and tablets can spur privacy-related discussions and initiatives, then, so I thought, an innovative technology such as Google Glass that puts cameras on people's faces and allows recording without notice would also spark value-laden discussions.

My presupposition was that, in the case of Google Glass, multiple meanings of privacy could exist that could be made visible through different modes of appropriation. This inspired me to conduct a preliminary empirical exploration of Google Glass's appropriation as specifically connected to the value of privacy. Such an exploration would not serve as a case study analysis, where the collected data has a thorough methodological grounding and is explained through a certain theoretical prism to yield conclusions. Rather, I envisioned the opposite: I wished to observe the phenomenon of moral hermeneutics with a mix of methodologies in mind and determine which questions and challenges arise in relation to it. It is in this sense that the exploratory study that follows is not a classical case study: instead of receiving answers to preset questions, I want to understand what questions I should ask further to explore the phenomenon of moral hermeneutics.

Inspired by the webcam sticker example, I concluded that the most interesting observations originate in the daily lives of people. However, when I originally became interested in this subject, in the beginning of 2015, Google Glass was not a widespread technology and was available only to few people, predominantly in the United States, who were ready to pay a large sum to test it. This presented a practical challenge to my preliminary empirical exploration, if I wanted to talk to people in real life. Already in 2015, however, the ability to discuss anything online with anyone anywhere on the globe was widespread. Particularly regarding new technologies, YouTube has frequently been used for reviewing new gadgets and receiving quick, diverse reactions from audiences in the form of comments. I decided that for an open empirical test bed, a digital ethnographic study of YouTube comments, coupled with the analysis of how people appropriate Google Glass, could be a pragmatic method to shed light on the connection between appropriation and moral hermeneutics.

Regardless of the exploratory nature of the study, the step from intuition to practice requires respect for the participants of the study and a transparent manner of conducting the investigation. The following section presents the setup of the study and describes the study itself in more detail to demonstrate how Google Glass implicitly co-defined the value of privacy through the process of appropriation.

TESTING THE ASSUMPTIONS OF TECHNOLOGICAL APPROPRIATION: GOOGLE GLASS AND THE VALUE OF PRIVACY[1]

Although mixed-reality goggles are not yet widespread, privacy-related concerns about them already exist. When Google introduced Glass in 2013, some businesses declared their spaces a "Glass-free zone," concerned that the embedded video camera compromised their clients' privacy. Glass augments human perception by providing an additional layer of information that blurs the boundary between the public and the private in new ways. In doing so, it further challenges the already messy endeavor of attempting to make sense of privacy in the digital age (e.g., Regan, 2002; Solove, 2002). The technology had a thorny path to the market: Google withdrew Glass for redesign in 2015, and in 2017, it introduced an updated device for enterprise use, continuing the work on Glass for mass consumers (Levy, 2017). However, mixed-reality glasses such as HoloLens (Microsoft Corporation, 2015) and Spectacles (2017) by Snap (formerly Snapchat) have recently entered the market, differing from Glass in intended uses but resembling it in having embedded cameras. This keeps the privacy discussion regarding Google Glass relevant: before technologies similar to Glass become widespread, it is necessary to understand why people point out privacy in the presence of these technologies. This makes Glass a fitting example for my pursuits of exploring the appropriation of technologies and their relation to moral hermeneutics. In an empirical study of online discussions that follows, I investigate how the notion of privacy used for the moral evaluation of Glass is implicitly redefined in interactions with the anticipated and actual mediating roles of this technology in human experiences and practices.

The value of privacy frequently appears in public debate and policy-making, but despite its dominant legal and corporate definition as information control and management (e.g., European Parliament, 2002; Google, 2013), a unified generic meaning of privacy has not been developed. Historical analyses demonstrate how the introduction of new technologies has gradually changed the meaning and practice of privacy (e.g., Mayer-Schönberger, 2009; Shapiro, 1998; Solove, 2002). Moreover, Steijn and Vedder (2015) have demonstrated how conceptions of privacy vary among different age groups; because the concerns and vulnerabilities of people differ in every life stage, the young and the elderly have different interpretations of privacy.

To study people's appropriation of Glass, I will utilize the blend of empirical philosophy, building primarily upon the ethnographic method of Mol (2002). According to this method, different practicalities enact different configurations of what a value means, leading to an ontological multiplicity that Mol calls the "body multiple." In this first attempt to study technological appropriation, I wish to connect the ontological multiplicity of values to the mediating roles of technologies: how are specific accounts of privacy articulated in connection to the specific ways in which technologies co-shape practices and experiences? To understand the privacy implications of Glass and meaningfully engage with them, I follow this technology through the practices it produces. To do this, I investigate a YouTube video on how to use Google Glass and, more specifically, the manner in which people reflect on Glass in view of their lives and understanding of privacy.

Google and Glass: "Back in Control of Your Technology"

Because corporate discourse co-shapes users' perception of technologies, I first examine how Google positioned Glass and how it discussed privacy. According to Glass's website, "Our vision behind Glass is to put you back in control of your technology" (Wayback Machine, 2015), which one can achieve through instant search and updates, picture or video recording (even on the blink of an eye, literally [Google, 2015]), and sharing information. Everything captured with Glass is accessible at any time due to continuous synchronization with Google Cloud. Google envisions Glass users as proactive individuals in control of their lives, activities, and information.

Being in control of information is also the primary principle behind Glass's security and privacy policy (Google, 2013), which highlights that although all Glass recordings are automatically backed up in Google Cloud, it is the user who decides with whom to share them. Concerning non-users, Google built in "explicit signals" to notify people nearby when Glass is recording, through screen illumination, a red light, and the use of voice commands, and called on the best judgment of Glass users when recording (Google, 2015). However, data protection authorities worldwide criticized the insufficiency of these signals as well as the lack of technical information available regarding how Google handles the data collected by Glass (Office of the Privacy Commissioner of Canada, 2013).

In 2014, Google introduced an "etiquette" guide for Glass Explorers, the first test users of Glass, designed to clarify appropriate usage, consisting of a short list of "do's and don'ts" to help Explorers adopt "collective wisdom" (Google, 2014) regarding using the device in social settings. Some of the "do's" suggested sharing captured experiences on social networks and interacting with Glass via voice. One notable suggestion was requesting permission from people when recording them, highlighting that Glass does not differ from a smartphone regarding camera use. This suggestion was reiterated in the "don'ts" as "[Don't] be creepy or rude (aka, a 'Glasshole')" (Google, 2014), asking Explorers to respect the privacy of others and apply rules regarding smartphone cameras to Glass. According to the etiquette guidelines, "Breaking the rules or being rude will not get businesses excited about Glass and will ruin it for other Explorers" (ibid.). Google's Glass etiquette essentially suggested adapting the conventional social rules to Glass. For instance, a notable "don't" was "[Don't] Glass-out," arguing against focusing on Glass for extended periods of time and for adjusting to social situations, even if this requires taking Glass off. The etiquette guidelines attempted to address an emerging pattern of socially contested behavior by Glass wearers and trust the better judgment of Explorers, asking them to "use common sense" (Google, 2014).

Users and media agencies preempted Google's initiative. I examined the first-of-its-kind Glass etiquette by Mashable, an online technology-review platform. A 1 min 46 sec video depicts in a satirical manner why some refer to Glass users as "Glassholes" and how to avoid being one (Mashable, 2013). Provocative scenarios present inappropriate uses of Glass—during a date or in the toilet, consulting search engines during conversations, and so on. The video engages viewers in reflection, thus presenting an interesting object of study for the preliminary empirical exploration of technologically mediated morality. The video went viral after its release on May 16, 2013, generating 1,434,719 views and 2,064 comments, all of which have been processed for this study.

YouTube, a social media platform with user-generated video content, invites open discussions about content and any topic provoked by it (Chenail, 2011). Although videos are staged interactions to which commenters react, the free choice of language, style, and expression allows commenters to engage on their own terms (Veen et al., 2011). This appropriation study, albeit the first attempt, purports to be a form of digital ethnography. As such, it still requires following ethical guidelines. Besides obtaining approval from the Ethics Committee at the University of Twente for this study, I have followed the recommendations of Markham and Buchanan (2012) and Hewson and Buchanan (2013) on ethical decision-making in Internet research. The public nature of YouTube comments does not require registration to access them. I anonymized the names of the commenters (e.g., *Commenter 1*) and removed any identifying information, such as the date, time, or location of posting. The original spelling stands.

I collected the comments manually and analyzed them using MS Word. Focusing on comments concerned with Glass-related uses while discarding promotional statements, incomprehensible symbols, and short expressions (e.g., "+Like") allowed me to narrow the original 2,064 comments[2] to 96, which formed the basis of an in-depth analysis. I used coding and thematic analysis as the elements of discourse analysis (Fairclough, 2013) and grounded theory (Charmaz, 2006) to analyze the narrative of the commenters and approach the data systematically. This allowed me to explore how commenters use contingent normative evaluations on Glass, particularly concerning the value of privacy, and how the commenters positioned the privacy discussions in their environment and in relation to Google. To qualify as a theme, a shared matter of concern must appear in at least ten separate instances. This explorative study also presents idiographic sensibility by considering equally relevant both single comments not fitting overarching patterns and comments that can be thematized (Smiths, Flowers & Larkin, 2009, pp. 37–39).

The complex narrative of the comments and my idiographic commitment has enabled me to produce rich findings, deepening my understanding of how people appropriate new technologies such as Glass. The qualitative study of YouTube comments provides a snapshot into privacy discussions in relation to Glass, indicating certain trends in privacy formulation "at one place in time" (Potts, 2015). As such, the results of this study do not pretend to be representative but rather are of an explorative nature. Therefore, I invite readers to approach the study as a suggestive illustration of how people reason with new technologies.

Below, I first present and critically reflecton the multiple interpretations of privacy that emerged in the YouTube discussions. In interpreting the YouTube narrative about Glass, I examine the nature of the practices that commenters describe, the primary issues at stake, and the values at play. I then inquire into why and how privacy is important for each practice and how people perceive and envision specific mediations of privacy by Glass. Finally, I make an inventory of the ways in which the value of privacy is implicitly articulated and defined.

Reasoning with Privacy

I first explore how and in which contexts commenters refer to privacy. A significant privacy-related discussion present throughout the comments concerned the fear that Google cooperates with international government structures to collect, store, analyze, and share large amounts of private information of Glass users and of any bystanders in their recordings.

Excerpt 1:
Commenter 1

1. You must be stupid to buy this. Putting your whole life and privacy
2. in the hands of a personal data-hungry company like Google.

Commenter 2 in reply to Commenter 1

3. Get used to it, Facebook, and even YouTube has your private information
4. (Google is YouTube). If you're really that paranoid then don't do a half . . . job,
5. abandon the internet completely.

This excerpt illustrates how privacy appears as a black-and-white argument to either use Glass and accept the supposed loss of privacy or abandon using it to preserve privacy. The privacy consequences of Glass are presented as self-evident, undeniable, and impossible to mitigate. Thus, the context within which the privacy discussion involving Glass emerged concerned the lack of transparency on how Google aggregates and treats the data collected by Glass.

The analysis of sociomaterial practices as presented by commenters online revealed a rich and complex narrative about privacy as a value. Commenters discussed privacy as a limited access to the self ("Addressing the GlassHole onslaught"), the privacy of personhood, the privacy of communication, privacy in public places ("You should be on guard!"), as well as privacy in relation to experience and memories, identity building, activity, and the control of information ("The end of privacy as we know it"). Below, due to space limitations, I present four of these privacy conceptions, accompanied by a mediation analysis.

1) *Privacy of communication: "Nail in the coffin of social grace"*

Excerpt 2:
Commenter 3

1. Wearable Internet is certainly the future, and probably the nail in the coffin of social grace.

Commenter 4

2. Not everyone is okay with the idea of a camera constantly being pointed at his or her face.
3. In fact wearing Google Glass on a date should be a definite no-no as they can make your date feel
4. uncomfortable and uncertain about what is going on behind that device.

Commenter 5

5. [W]ho wants to guess if you are really paying attention or reading a text.
6. You will be more interested in icons floating across your field of vision than talking
7. one on one. Recording me talk? Taking photos? Who knows what you're doing.

Commenter 6

8. There is absolutely etiquette for glass. I'm from a big city [. . .] where individuality thrives
9. but here in the good 'ole south [. . .] conservatism goes a long way.
10. That being said, I have vigilantly conscious when and where to wear glass.
11. There is an evolving glass etiquette as we speak.

Excerpt 2 suggests that Glass can mediate a set of practices related to everyday communication. The commenters appropriate Glass as an element of suspicion during interpersonal communication, leaving the other party "to guess if you are really paying attention" (line 5) or "taking photos" (line 7), and even framing Glass as "the nail in the coffin of social grace" (line 1). Excerpt 2 represents the widespread assumption that Glass users will violate tacit social norms. However, as *Commenter 6* suggests in line with Van de Poel and Kudina (2022), social etiquette co-evolves with the introduction of new technologies, confronting existing norms of behavior with new technological practices. Nonetheless, cultural and social landscapes are fundamental in navigating new technologies, or as *Commenter 6* put it, "I [am] vigilantly conscious when and where to wear glass" (lines 10–11).

Privacy and attention are necessary conditions to foster interpersonal relations and express identity appropriate to a certain social context (Solove, 2002). As Excerpt 2 indicates, Glass challenges these conditions by presenting the possibility of being constantly watched without knowing whether you are being recorded and by forcing the interlocutor to guess what the Glass user is really doing. The design of Glass both suggests certain use practices (i.e., conducting several social activities simultaneously) and co-shapes how users can achieve them. Glass is positioned above the user's right eye, in the direct field of vision, "to cater to *microinteractions*, allowing the wearer to utilize technology while not being taken out of the moment" (Firstenberg & Salas, 2014, p. 11). However, using Glass requires concentration on the screen and close attention to frequent visual notifications and navigational aural cues, in addition to interaction via voice commands and by tapping the device. In practice, this requires Glass wearers to often focus on and interact with the device itself, which complicates interactions with other people (Honan, 2013; Koelle, Kranx & Moller, 2015).

Overall, Excerpt 2 suggests a transformative effect of Glass on communication practices because it mediates attention and focus, values constitutive for the privacy of communication. Following one commenter, human norms of interaction co-evolve with new technologies, suggesting that with time, Glass will not only mediate what such norms are but also what meaningful communication constitutes.

2) *Privacy as limited access to the self: Addressing "GlassHole onslaught"*

Excerpt 3:
Commenter 7

1. I'm sorry, those who pull these kinds of stunts would more than likely get their snotbox
2. busted by someone who isn't cool with it. Google glass with caution. I'm just sayin'.

Commenter 8

3. I don't want to be in the sauna at the gym and have some GlassHole walk in.
4. I remember how irritating it felt in 1990 when some self-important person with
5. a Motorola Brick would decide to call someone while waiting in line at the grocery.
6. The GlassHole onslaught: 50x as intrusive.

Commenter 9

7. If you point those things at me or a member of my family and record footage for the NSA
8. you will find those glasses shoved up your glasshole.

In Excerpt 3, Glass appears as a mediating boundary object between what commenters consider inherently private, even in the most public places, and what is violated when the device is introduced. *Commenter 8* worries about Glass users violating his or her bodily privacy and the involved sense of dignity, illustrated by the retrospective cell phone example (lines 4–6). Providing recognition of a contextual use of Glass (Steeves & Regan, 2014), she or he engages in a negotiation of the public-private spheres with Glass as an active boundary object. Curiously, by relating to his or her own feelings about someone using a phone in public, *Commenter 8* evokes an idea of moral hermeneutics by depicting how human understanding of appropriate behavior has changed with the introduction of cellphones, or more generally, how technologies mediate moral frameworks. The perceived mediation of Glass concerns an undesirable intrusion into certain spaces. Anticipating public backlash concerning the video recording Glass, *Commenter 7* similarly suggests using Glass proportionally to the context (line 2) without specifying what such Glass etiquette would entail. Comments here demonstrate how the introduction of Glass potentially destabilizes existing norms and how to reflect upon this through deliberation and comparison. Other commenters, represented by *Commenter 9*, suggested less formative ways to reason with Glass. Some understand Glass as a direct threat to the privacy and security of themselves and their loved ones (lines 7–8), threatening its users with sabotage and physical injury.

Excerpt 3 displays an intricate web of values in relation to Glass, such as proportionality, fairness, responsibility to protect their loved ones, justice, and accountability. Together, they conjure an understanding of privacy as the desire for limited access to the self and indicate the multidimensional nature of privacy.

3) *Privacy of experience and memories: "Sharing some things [is] fine but why everything?"*

Excerpt 4:
Commenter 10

1. How about going dirtbiking. . .and *not* showing it to the entire internet?
2. Just enjoy your life.

Commenter 11

3. God I hope Glass Fails. . . . Does anyone remember or value real experience?
4. Or memories?. . . . [S]haring some things are fine but why everything?. . . .

Presented with an option to easily record one's surroundings through Glass, coupled with Google's encouragement of users to post their experiences online, Glass users share recordings of their most mundane to most exciting experiences. Although it is one's choice whether to watch such videos, the multitude of Glass recordings online and the behavior-mediating design of the media platforms presenting these videos (e.g., activated by the default option to "autoplay" the next clip) intensify human curiosity and diffuse the criteria for decision-making.

Excerpt 4 suggests that the privacy of remembering and, mirroring the concern, the privacy of forgetting, are at stake with Glass. Frustrated with the extensive sharing of personal content online, *Commenters 10* and *11* believe that a frequent sharing of personal experiences prevents one from enjoying the present (line 2) and takes a toll on the value of such experiences (line 3). I interpret their frustration as a desire to reclaim the right to form good memories. Mayer-Schönberger (2009) endorses the right to be forgotten in the digital age as a legal mechanism for dealing with the mediating impact of online sharing and storing practices. He discusses the case of a teacher who was fired because of images on her Facebook page portraying her with alcoholic beverages. This example illustrates the repercussions of a collision "when actions that are normatively appropriate in one context are revealed to members of another audience where norms are different" (Blank, Bolsover & Dubois, 2014, p. 6). However, the human ability to forget, mediated by the immeasurable capacity of the Internet to remember and coupled with diverse self-representation online enabled by Glass, presents a favorable background for conflict-laden situations.

Overall, the commenters in the excerpt discuss the overexposure and presence on the Internet that Glass enables. This has allowed me to discern privacy in the context of experience and memories, with the accompanying intricate interplay of values such as proportionality, balance, appropriateness, and choice, as well as remembering, forgetting, and balancing normative expectations.

4) *Privacy in the public space: "You should be on your guard"*

Excerpt 5:
Commenter 12

1. These will end up being abused by the police and government so damn much,
2. the end of privacy as we know it. Plus everything you do and say will be recorded
3. in public places now, its scary to even think about.

Commenter 13

4. this should be prohibited . . . every[one] can take pictures and videos from me,
5. everywhere in the public space.

Commenter 14 (in reply to Commenter 13)

6. Because there is an expectation of privacy out in public right?

Commenter 15

7. Lack of privacy comes in many flavors . . . .
8. There's the—oncoming tidal wave of CCTVs in public spaces—universal behavior
9. of anyone with a phone feeling that it's OK to take pictures wherever—

10. [. . .] So now we have people who can take your picture while non-surreptitiously
11. (you should be on your guard when addressing someone you don't know
12. who is wearing Glass) facing you.

Following the commenters in this excerpt, Glass mediates the value of trust in bystanders and enhances the curiosity of bystanders by enabling users to—randomly or not—record them and use the recordings upon one's best judgement. Commenters appropriate Glass as an abuse of privacy in public, be it in dystopic undertones—"the end of privacy as we know it" (line 2)—or with irony—"Because there is an expectation of privacy out in public right?" (line 6). Such anticipation joins the fears of Google cooperating with governments and law enforcement for the purpose of policing (line 1). The shared assumption is that no room for anonymity exists where Glass monitors, inspects, and singles out.

Highlighting the disclosed observation practices that Glass enables, *Commenter 15* lamented the "lack of privacy" (line 7). The ambiguity as to the purpose, extent, and context of recording with Glass challenges the practices of the development and representation of the self in public. What distinguishes Glass from public CCTV surveillance is that with CCTV, due to security reasons, it is necessary to focus on single individuals, and recorded data is managed in accordance with the legal requirements of intent and proportionality (Taylor, 2002). While recording with smartphones does not manifest the intentions of the users, it does make the action of recording visible and/or audible. Glass users, however, neither visibly nor audibly manifest their intentions. The warning of *Commenter 15*—"You should be on your guard" (line 11)—mirrors the conclusions of Koelle, Kranz, and Möller (2015), suggesting that in the absence of any signals, people assume they are being recorded when faced with devices such as Glass.

Excerpt 5 represents deliberations on the expectation of privacy in the public space and the nature of such expectations in the age of recording devices. Regardless of the open and shared aspects of the public space, an expectation of privacy is inherent to it as an enabling condition for contextual self-development and disclosure (Roessler & Mokrosinska, 2013). Such privacy as civil inattention highlights the social dimension of privacy, "when respect and reserve are displayed towards others" (ibid., 782). Similarly, Tonkiss (2003) suggests that an ethics of indifference is a necessary condition for coexistence in the public space, stemming from "'side-by-side' relations of anonymity" (298) and an ethics of "look[ing] past a face" (301). Privacy as civil inattention, which enables sociality and representation in public, hinges on the civil indifference of others, the condition that, according to Excerpt 5, Google Glass removes.

Reflecting on the Preliminary Study

The mediation analysis of the YouTube comments above demonstrates how moral hermeneutics has accompanied the introduction of Glass. In particular, following van de Poel and Kudina (2022), it depicts a case of value dynamism, when confronted with a morally problematic situation, the value frameworks undergo reconceptualization and reinterpretation to make them fit to address the new problematic situations, induced by technologies. This appropriation study tentatively illustrates how the introduction of Glass might mediate the social practice of communication, the responsibility and proportionality of using Glass in public and private encounters, and the relation of Glass to memory making and to maintaining the expectation of privacy in public places. The study suggests how people anticipate the mediating role of Glass in their daily experiences and practices and how, in connection to this, specific articulations of privacy become visible. The technological mediation approach does not provide generalizing predictions on the possible societal or normative impact of Google Glass, nor does it apply static normative conceptions to approach the device. Rather, it draws on specific human practices and experiences to identify the appropriation strategies people employ to make Glass fit into their daily lives, enabling the rearticulation of normative concerns.

Personal stories in the comment threads demonstrate the multiplicity of what privacy means in the specific practices enabled by Glass. For instance, the same set of issues, such as the questionable trustworthiness of Glass users' behavior or uncertainty regarding the boundary between public and private, foster different practices that, depending on the context, conjure up different understandings of privacy and why people appeal to it. At the same time, it is interesting to see how YouTube commenters perceive privacy as primarily related to surveillance concerns rather than exploring its local meaning. This could be explained by promotional activities by Google to amplify the outdoor use of the device and its sharing practices. By identifying specific issues that become heightened with the introduction of Glass, such as personal freedom and well-being regarding the questionable behavior of Glass users, commenters shape the understanding of privacy that is meant to safeguard these practices. This study thus illustrates not only that, but also how sociomaterial practices influence how people interpret privacy in the context of Glass appropriation.

Lastly, this first appropriation study suggests that people approach privacy not as an ephemeral entity but rather as something that originates and becomes embedded in specific sociotechnical practices. It showcases how people, confronted with Glass, expose the tacit understandings of privacy they previously possessed, review them, and contest the suggested definition of privacy as the control of information, thus allowing new value interpretations that work in the Glass-enabled practices. Viewed from the pragmatist angle of values as related to practices, the value of privacy appears as an appeal, a working solution to the new situations that Glass enables, revealing different dimensions according to the situation at hand and the specific concerns of various individuals. The value of privacy, as traced and analyzed in this preliminary investigation, is not static but dynamic. On the one hand, it is generic and sufficiently universal to cover a multitude of different practices, and on the other, it is flexible enough to reveal corresponding dimensions sensitive to the exigencies of specific situations. The technological mediation approach has allowed an exploration of several facets of the value of privacy, mediated by Glass, while being grounded in specific human practices and concerns.

TOWARD A COMPREHENSIVE STUDY OF MORAL HERMENEUTICS

This chapter has set out to explore the fitness of the concept of human appropriation of technologies as a way to study morality-in-the-making, the basic infrastructure of moral hermeneutics. The conceptual clarifications allowed positioning the reinterpretation of values in the process of appropriation, the sense-making of people regarding specific technologies, and fitting them into the accumulated interpretation schemes. Beyond presenting the conceptual explorations, this chapter presents a preliminary investigation of how the moral hermeneutics can become manifest through the study of technological appropriation, through the example of Google Glass and privacy. I envisioned such an exploration as a thermometer to test an intuition about moral hermeneutics (i.e., that it is possible to observe it empirically), which could help to direct further theoretical and empirical exploration. In parallel, I wanted to understand the potential challenges that could surface in exploring the moral hermeneutics as related to technologies to eventually produce a theoretically and empirically sound way to do so.

The preliminary appropriation study of Google Glass has shown that what people mean by the value of privacy changes in relation to this technology. A connection does indeed exist between technologies and values, whereby values are not stable backgrounds but are flexible and responsive to the sociotechnical practice at hand. Such observations, however preliminary, push the boundaries of the technological mediation account further, for they show that moral mediation includes dynamism in the value frameworks. More specifically, the preliminary study of Google Glass illustrates a challenge in explaining how moral hermeneutics occurs, both conceptually and empirically. What does it imply for value theory? What does it imply for technological mediation and appropriation?

Considering the empirical method, how can one more closely study the interaction between technological mediations and the meaning of values in a methodologically thorough intertwinement of empirical and philosophical analysis? As the Google Glass study succinctly illustrates, there is a deep hermeneutic dimension both at the level of commenters making sense of Google Glass and of analysis, where I, as a researcher, interpret the interpretations of the commenters. Moreover, the value of privacy is not always explicitly mentioned by the commenters; rather, I interpret certain situations through the prism of privacy based on the accompanying concerns and context. In short, the method used to study the moral hermeneutics and appropriation of technologies should account for the dynamics of interpretation on different levels, including the role of researcher. Finally, demonstrating how the meaning of the value of privacy co-evolves with Google Glass begs the question of how to still do the ethics of technology when the values that guide people in design and evaluation change in relation to the discussed technologies.

As the reader may notice, these questions mirror the research lines defined in the preliminary theoretical exploration of the introduction: expanding the technological mediation account with moral hermeneutics, developing an account of appropriation, designing and implementing a research methodology to study moral hermeneutics empirically, and correlating the findings with the broader field of the ethics of technology. The first appropriation study of Google Glass in this chapter serves as proof of principle that one can study and analyze the phenomenon of moral hermeneutics while also helping to substantiate and better focus the research lines, as it grounds them in the messy reality of exploring moral hermeneutics, albeit in a tentative manner and without a guiding methodology.

The preliminary investigation of Google Glass and privacy is in line with the theoretical explorations of morality as an ecosystem and the pragmatist accounts of values and value change developed in the previous chapters. However, it still points to a number of theoretical and empirical assumptions that must be clarified to construct an account of moral hermeneutics. In what follows, I must (1) develop a thorough methodology to empirically scrutinize moral hermeneutics (chapter 4) and (2) develop an encompassing principle of moral sense-making that is hospitable to the ideas of value change induced by technologies (chapter 5). The hermeneutic dimension of mediation must become more concrete to explore value dynamism and change, and showing how people-appropriate technologies can help to achieve this. I proceed to such elaborations in the chapters that follow, beginning with a quest for finding a fitting method to study moral hermeneutics through technological appropriation.

NOTES

1. A modified version of this section appeared in the following open-access article: Kudina, O., and P.-P. Verbeek. (2019). Ethics from within: Google Glass, the Collingridge dilemma, and the mediated value of privacy. *Science, Technology, & Human Values, 44*(2), pp. 291-314.

2. In April–June 2014, during the empirical stage for this study, the number of comments below the video was 2,064. However, during the review of this study in 2022, the number of comments below the same video decreased to 589. A possible explanation could be a recently enhanced filtering policy of YouTube, where human and AI-based assistants remove the content (also comments) containing spam, hate speech, etc. (https://support.google.com/youtube/topic/2676378?hl=en). Many of the original comments indeed contained spam and hate speech, which I filtered manually. The ninety-six comments taken for a close analysis remain intact on the site as of March 2022.

Chapter 4

Interpretative Phenomenological Analysis as a Method to Study Moral Hermeneutics

A QUEST FOR A METHOD FOR EMPIRICALLY PHILOSOPHICAL INVESTIGATION OF MORAL HERMENEUTICS

While the theoretical elaborations in the introduction suggest the necessity of a consistent empirical method to study moral hermeneutics, the Google Glass case from the previous chapter vividly illustrates this need. As demonstrated in the online discussion analysis, producing a nuanced view of how values undergo reconceptualization in an encounter with technologies requires a new method of conducting empirical philosophy. From here stems the importance of having a well-rounded empirical methodology, one that captures both the theoretical and empirical hermeneutic process of value interpretation. As demonstrated in the comment analysis, existing qualitative accounts, such as discourse analysis and content analysis, with elements of grounded theory, allow for an understanding of what commenters say at a given time. In the study of the YouTube comments, these methods provide a glimpse into how people appropriate and bestow meaning upon technology. However, this barely scratches the surface of interpretation, excluding deeper meaning-making connections in relation to the individual lives of the participants as well as the researcher's involvement in this. Particularly considering the need to highlight the hermeneutic dimension of value making, an empirical method to examine moral hermeneutics must exceed the immediate content and consider not only *what* people say but also *how* they do so.

Additionally, a method to study how people appropriate technologies must be able to identify how people incorporate the technology in question into their daily lives and interpretative frameworks, or how they attribute it with meaning. It must simultaneously demonstrate how, during the appropriation process, certain normative concerns crystallize, whereby the entangled norms and values are confirmed, challenged, or develop new facets. For instance, the exploratory appropriation study conducted in chapter 3 reflects only a fraction of possible moral deliberation regarding new technologies. The study was limited by the static nature of the YouTube comments (once captured for analysis), concealing much of the meaning-making process leading to the writing of these comments and presenting only their solidified expression in text. Intuitively, it would be worthwhile to explore how people make sense of new technologies in their conversations, or rather, in their lived dynamic language, rather than through textual interaction. Moreover, text comments are not always available, nor are the developing versions of technologies, as with the Explorer version of Glass. It is safe to conclude that the empirical study of moral hermeneutics I aim for must exceed the static interactions at a single point in time offered by online comment analysis. Thus, a method for an appropriation study and moral hermeneutics must not only provide insight into the dynamism of the moral landscape of people, enabled by the technology in question, but also be able to reflect the dynamism of human sense-making activities. While qualitative empirical methods are the most appropriate to reflect the postphenomenological focus on human experiences and sense-making practices, the challenge is to choose a suitable method among the many that exist in this field.

The method of the study co-shapes its process and results and must therefore share at least some of the theoretical foundations of the study. Commenting on the choice of a method for a study, Smith, Flowers, and Larkin (2009) note that "this is not so much a matter of choosing 'the tool for the job' [. . .], but a question of identifying 'what the job *is*'" (p. 43, original emphasis). Because an appropriation study is hermeneutic in nature, hermeneutic philosophy can provide insight regarding the search for suitable methods. In particular, Gadamer's emphasis on the intimate link between language and conversation, on the one hand, and the process of interpretation, on the other, can serve as a starting point.

In Gadamer's hermeneutics (2006), language is critical for understanding because reality becomes intelligible to us through language, where it functions as a medium or a lens that sharpens the perception of reality and brings it into focus. It is through speaking with others or engaging in a mental conversation with oneself that a particular meaning comes into being. At the same time, language, belonging to the effective history of the interpreter, not only mirrors reality but also necessarily co-produces it, expanding and distorting ways of perceiving something. As such, the use of language in attempting to interpret reality is not neutral. If we follow Gadamer's ideas in a search for a method for moral hermeneutics, then to understand how one reaches an understanding, how the act of interpretation occurs, necessitates reflecting on the use of one's language and accounting for its mediating properties in the course of a conversation.

Attention to language does not presuppose performing a thorough linguistic analysis but rather prompts one to be aware of how a particular linguistic tradition actively participates in the sense-making process. This requires a focus primarily on what is being said and in which context, what metaphors and comparisons are being invoked to communicate meaning and the emotional tone of the speaker. Reflecting on the conversation in this manner allows the nuances in the sense-making activity to be contextualized and brought to the fore, which is essential for the study of the human appropriation of technologies.

Verbeek (2015), when commenting on possible empirical methods to capture how people appropriate technologies, suggested that the method of Conversation Analysis (Sacks, 1992; Edwards & Potter, 1992; Te Molder & Potter, 2005) can help capture how, through conversations, people construct a world around them and the specific meaning of technologies in it. Accounting for language and conversation insights, the broader method of Conversation Analysis and Discursive Psychology (CA&DP) is thus a possible candidate for an empirical method for an appropriation study. It was developed by Derek Edwards and Jonathan Potter (1992) to methodologically study conversation and its implications for social interactions and daily life. Another candidate for the empirically philosophical exploration of moral hermeneutics and technological appropriation is the method of Interpretative Phenomenological Analysis (Smith, Flowers & Larkin, 2009), used in psychology to understand the lived realities of people and their sense-making practices as well as originating in the field of philosophical hermeneutics. The following section examines both methods in detail as well as questions their fitness to study moral sense-making in relation to technologies.

THE METHODS OF CA&DP, AND INTERPRETATIVE PHENOMENOLOGICAL ANALYSIS

Exploring the CA&DP Method

The method of CA&DP reveals the morality of everyday life through its focus on how people talk and non-verbally interact through the sequential and rhetorical analysis of a conversation (Sacks, 1992; Edwards & Potter, 1992; Potter, 1996; Te Molder & Potter, 2005; Te Molder, 2008). Based at the intersection of sociology, ethnomethodology, linguistics, and social psychology, the method emphasizes the interaction patterns that underlie human conversations (Hutchby & Wooffitt, 2008). It demonstrates how such patterns shape the legitimacy of discussion points as well as how, through these interaction patterns, people attribute to themselves and distribute to others epistemic rights and responsibilities.

Human talk is perceived as action-oriented, where participants, with each utterance in a conversation, perform certain actions (e.g., attributing responsibility or praise). As such, *how* people talk—how they take turns speaking, which words or utterances they use, the tone of their voice, and their pauses and sighs—are considered by the CA&DP method as tools for achieving specific goals in an interaction. In this sense, human talk is not only neutral but also deeply normative because it involves issues of identity and responsibility.

At its core, the CA&DP method explores the human entitlement to speak. For instance, turn taking in a conversation, agreement on the distribution of roles, and the determination of who is accountable for attributing meaning and interpretation to the matter at hand all shape specific interaction patterns in the course of a conversation. The CA&DP method reflects on how such interaction patterns constitute and are constituted by the object of the conversation, suggesting that both a conversation and the knowledge at its outcome are deeply normative enterprises (Myers, 2004).

More specifically, the CA&DP method aims to recognize the formative and moralizing nature of interaction patterns in a conversation. This goal is closely linked with recognizing and reflecting on specific epistemic rights and responsibilities in a conversation that are concerned with the actual and expected knowledge of the speaker and the other parties in the conversation (Haen et al., 2015). The CA&DP method explores how such epistemic rights and responsibilities co-shape a conversation and how participants attribute such rights to themselves and others, contest them, or agree with them. This involves the retrospective and prospective attribution of guilt, blame, praise, accountability, and obligation with regard to the object of the conversation (Heritage & Raymond, 2005).

As Haen and colleagues (2015, p. 167) note, epistemic rights and responsibilities closely relate to the identity of a person and their entitlement to speak, which ultimately influence the actions of people. The CA&DP method approaches a conversation sequentially, examining how participants in the conversation treat what is being said. Furthermore, it explores how alternative interpretations of reality (or of a specific object of conversation) are produced to counter those produced by other participants. The areas of interest here concern determining who is justified in claiming certain knowledge, how people hold each other accountable, how agreement or dissent regarding the distribution of roles occurs in a conversation, and other manners through which people coordinate the interaction process (Te Molder, 2008). Reflecting on such interaction patterns reveals the implicit morality of a conversation and points to the non-neutrality of the consequently produced knowledge claims.

In this regard, empirical material for reflecting on the nature of a conversation includes naturally occurring interactions in the ordinary context of human lives. As opposed to prearranged interactions—experimentally designed and set up (e.g., in interviews)—naturally occurring interactions present a richer and more untainted background to study "talk-in-interaction," "how sequences of actions are generated" (Hutchby & Wooffitt, 2008, p. 12), or how a conversation is achieved in practice. To this end, CA&DP analysts record the naturally occurring interactions and thoroughly transcribe them using a special transcription system (e.g., indicating time gaps, pauses, overlapping utterances, concurrent speech, word cut-offs, animated tone, intonation shift, etc.). The analysts then perform a sequential analysis, or "a next-turn proof procedure" (ibid., p. 13), to understand the properties of a conversation, its order and its structure as a process of social accomplishment. As such, the CA&DP method is less concerned with what participants say or the descriptive content of the conversation. Rather, in this method, center stage belongs to the interactional nature and form of the conversation that specific words help to achieve (Te Molder & Potter, 2005).

So far, the CA&DP method suggests itself as a natural candidate for an appropriation study due to its focus on human talk and conversation, and the morality that is achieved through them. Another candidate method concerns the Interpretative Phenomenological Analysis, or IPA for short, which focuses on the content of the conversation rather than its form and the meaning-making processes underlying it. I will now briefly explain its focus.

Exploring the Interpretative Phenomenological Analysis Method

Originally emerging from the field of psychology in the 1990s, IPA was meant to contrast with the discursive methods of research that were dominant at the time. Discursive methods focused primarily on how research participants construct accounts of themselves and their experiences in everyday talk. The CA&DP is a notable example of the discursive method. According to Smith (2011a), the prevalence of discursive methods downplayed the necessity of studying the content of what is said. Thus, IPA developed to bring the sense-making process to the forefront of research, focusing on the content of how people attempt to attribute meaning to various phenomena.

IPA finds theoretical grounding in phenomenology, hermeneutics, and idiography[1] (Smith, 2011a, b). Phenomenologically, IPA focuses on the study of the lived, embodied experiences of people and draws heavily upon Gadamer's hermeneutics. Particularly his principle of the hermeneutic circle and the productive nature of biases in producing an understanding became important methodological considerations for IPA (Smith et al., 2009). Furthermore,

IPA is a hermeneutic activity because it acknowledges that it is impossible to directly engage with people's experiences. A researcher is faced with a situation where she must attempt to make sense of the experience of which the participant is trying to make sense. For these reasons, IPA scholars refer to the research process in IPA as double hermeneutics, which clearly demonstrates how and whence the presented analysis is derived. This requires a detailed, close-up analysis of the rich experiential narrative of participants, resulting in balanced patterns of convergence and divergence across the research cases. By doing this, IPA manifests its idiographic commitment.

The "father" of IPA, Jonathan Smith, outlined detailed criteria and recommendations for a good IPA study (e.g., 2011a, pp. 17–18, p. 24; 2011b), which should feature "depth of interpretation, sensitivity of analysis, the importance of particular utterances" (Smith, 2011b, p. 59) and demonstrate "what the data are, how the data were obtained, *and* what the data means" (ibid., p. 60, original emphasis). In-depth, semi-structured interviews support this strong idiographic commitment of IPA to the detailed personal accounts of participants. Conducting interviews with a consequent rigorous analysis is a cornerstone of IPA. As Smith (2011a) notes, a successful IPA analysis must have "interpretative flair" (p. 23), whereas a good IPA piece, in general, "needs to be plausible and persuasive in terms of evidence presented to support the claims made" (p. 23).

The scope of the research must, however, remain manageable given the intensity and depth of individual experiences. Here, IPA also draws on Gadamer to say that it favors a small research participant group with rich idiographic findings over a large group that would inevitably require significant simplification of data. Gadamer's philosophical hermeneutics favored in-depth study over generalizations based on large numbers because it allowed for situated knowledge regarding how a certain phenomenon manifests. According to Gadamer, "The individual case does not serve only to confirm a law from which practical predictions can be made. Its ideal is rather to understand the phenomenon itself in its unique and historical concreteness" (1975/2004, p. 4). The IPA method follows this idea, and the top IPA papers have relatively small study samples, ranging from one to ten participants (Smith et al., 2009).

An IPA analysis denotes the developing sensibility of discerning sense-making in action and interpreting it according to participants' lived experiences (i.e., practice-based) and proactive agency (i.e., projective sense-making). As such, it is a hermeneutic exercise of *at least* two layers: an IPA researcher interpreting the interpretation activity of the research participants (Smith et al., 2009). In parallel, the material setup of the research provides additional layers for the interpretation process. For instance, how an IPA researcher captures the thoughts of the research participant (e.g., recording an interview with a phone or a professional recorder) also co-shapes her process of analysis, often simplifying it (e.g., presenting captured speech for later transcription) while also complicating the process (e.g., leaving the researcher to guess the parts of the recording that were less audible or interrupted, always maintaining the sufficient battery level of the device, etc.). The next step, the transcription of the recording, is also a hermeneutic exercise in itself because it transforms the speech of the participant into readable words, often in a specific manner (e.g., describing the intonation, emotional tone, or special circumstances of the interview, etc.). For this reason, I suggest that IPA research is an exercise of *at least* double hermeneutics, where triple and further layers can be discerned upon inspection. However, the goal is not to discern and analyze such possible hermeneutic layers in any given IPA study; rather, the goal is to maintain critical awareness of the projective and material processes that inevitably underlie the process of interpretation.

As a hermeneutic study into the sense-making activity of people and as idiographic by nature, IPA includes a thorough research and analysis procedure to structure the research and create verifiable results for the public at large. An IPA study collects data in the form of semi-structured qualitative interviews from a relatively homogeneous group of research participants who are united by a certain shared experience or other shared features. Transcription proceeds verbatim, highlighting in the text instances when the intonation changes or the participant laughs or pauses for a long time. To systematize the process of analysis in IPA, Smith, Flowers, and Larkin (2009) suggest approaching it in six steps, each non-exhaustive, iterative, and mutually informative: reading, initial noting, developing emergent themes, seeking connections among emergent themes, moving to the next case, and seeking patterns among cases.

The IPA method searches for patterns of meaning that are shared by multiple participants, regardless of the particular nature of their lived experiences. Nonetheless, it is crucial to remain true to how participants express themselves, their choice of words, and the terms used. As such, the development of emergent descriptions and themes in IPA is different from the coding technique that is dominant in the social sciences. This also explains why the IPA method favors manual data analysis over data processing software (e.g., NVivo or Atlas.ti) (ibid.).

Being idiographic by nature, the results of an IPA analysis do not intend to be fully generalizable, which reflects the subjective nature of the analysis. The subjectivity in IPA analysis is the outcome of the double (triple, quadruple, etc.) hermeneutics underway in the study. It does not discredit the study because the IPA method acknowledges it and identifies how the subjectivity of analysis emerges throughout the study. This subjectivity is a systematic product of the dialogue between the different levels of interpretation and is open for the reader to verify through both the researcher's analysis and excerpts of the participants' interviews. The multiple hermeneutic layers in the IPA study are what makes it "a reflective engagement with the participant's account" (Smith et al., 2009, p. 80).

Two core processes constitute an IPA analysis: moving from the descriptive accounts of participants to interpretative pieces and gradually uniting particular patterns into shared themes across multiple participants while retaining the commitment to the participants' choice of words (Reid, Flowers, & Larkin, 2005). The iterative and inductive nature of analysis allows the two core processes to be implemented. It embodies the idea of the hermeneutic circle, whereby one can arrive at the whole only by pivoting between the parts. In IPA analysis, the researcher never considers a piece of the interview transcript in isolation but rather positions it against the overall transcript. This requires several rigorous readings of the interview transcript, usually beginning anew after completing the analysis. Such an iterative and inductive mode of analysis accounts for the details and nuances in the data and ensures that the final interpretative narrative remains true to particular lived experiences and mirrors their gradual unfolding.

Such an extensive introduction to the IPA method presents it as a hermeneutic tool to explore the sense-making activity of people based on their historical and social situatedness. It also acknowledges the co-shaping interpretative efforts of the IPA analyst, embedded in the sociomaterial environment. The focus of interpretative activity can range from specific life experiences to specific objects in the lives of people, accompanied by a rigorous method to approach the information provided by research participants. Coupled with the phenomenological and hermeneutic heritage that underlies the IPA method, it appears to be a good candidate for a method to study moral hermeneutics through the human appropriation of technologies. The following sub-section compares the methods of Conversation Analysis and Interpretative Phenomenological Analysis to determine their suitability for an appropriation study.

Comparing the Two Methods

This overview of the CA&DP and IPA methods, albeit non-exhaustive, allows the examination of their fitness for the study of moral hermeneutics through human appropriation of technologies. The CA&DP method is focused on the construction of human talk and how morality materializes in the implicit and explicit goals that people materialize in their conversations. Recently, it has been used to expand the technological mediation theory regarding the active role of talk in the constitution of reality and the technological objects through which it consequently produces specific moral perceptions and actions. For instance, de Boer and colleagues (2020) used the CA&DP method to show

how technologies in neuroscientific practice materialize in the conversations of the practitioners and how this fosters specific notions of "causality" and "reality," and thus guides the visual attention of neuroscientists in specific non-neutral ways. As inferred from Gadamer's hermeneutics (2006), the appropriation study also has a close affinity with language and primarily with sense-making activity. However, as I explain below, the specific focus of the CA&DP method on the interactional nature of conversation ultimately causes it to diverge from the goals of the appropriation study and the content-driven nature of the moral hermeneutics study.

Firstly, although it does focus on human conversations, the true focus of the CA&DP method is the morality of the conversation itself, or *how* people speak, instead of *what* they are saying: "The focus is not on individual cognitions (intentions, motives, attitudes) but on understanding how the talk is treated by others (as blame, compliment, et cetera)" (Haen et al., 2015, p. 168). The appropriation study, conversely, is concerned with the descriptions, opinions, attitudes, and biases of a specific individual rather than the interactional goals of a conversation. The content, as such, is of greater importance for the appropriation study than the form or structure of what is said.

Secondly, the method for the appropriation study need not limit itself to the naturally occurring interactions of people. Although sense-making can occur in groups, it is not restricted to them and can occur in individual encounters with a researcher, for instance, during interviews. Particularly because the object of the appropriation study is frequently a new or emerging technology that has not yet entered the market or has only limited market exposure, naturally occurring interactions either using the technology in question or discussing it would be rare to encounter. In such circumstances, fostering someone's thought processes, opinion making, and reflection on a new technology would require invoking their proactive agency by confronting the person with carefully designed information prompts, scenarios, or moral provocations. In this sense, staged interactions take precedence over naturally occurring ones, with open interviews (carefully designed to limit potential bias) providing a likely empirical background.

The CA&DP method uncovers the normative dimensions of the linguistic resources that people draw upon in conversation by analyzing how people say things, how they position themselves in a conversation, the types of words they use, and the turns they take while speaking. With this, the CA&DP method brilliantly depicts the morality of conversation, showing that it is, by far, a very normative rather than a neutral process. However, the depth and breadth of what is said remain concealed.

Additionally, as presented in chapters 1 and 2, the technological mediation approach has an interactionist and pragmatist take on values, where values do not exist in isolation but are enacted in the relation between people and their sociomaterial environment. Therefore, understanding the dynamics of moral hermeneutics is impossible without grasping the context that informs it and lies at the background of a particular human-technology relation. Moreover, appropriation is essentially a hermeneutic activity, whereby a person explicitly or implicitly interprets a new technology and attributes it with meaning. Here, every person possesses a richness of prior experiences, perceptions, and knowledge that informs their sense-making activity in an encounter with a new technology. It follows, then, that an effective method for studying the appropriation of technologies must maintain a focus on people, what they say and their larger context; the technology in question; and a sociocultural embedding. In this regard, the CA&DP, interested largely in attaining specific conversation goals, would give way to the IPA method, which explicitly focuses on the meaning-making processes and context.

In contrast to the Conversation Analysis method, IPA is not discursive but experiential by nature. It studies what people say "in order to learn about how they are *making sense* of their experience" (Smith, 2011a, p. 10, original emphasis). Because the focus of IPA is the participant's account itself, the transcription of IPA interviews—unlike that of the CA&DP method—does not require a detailed write-up of the prosodic elements of the conversation (all non-verbal cues or utterances, length of pauses, etc.) (Smith et al., 2009, p. 74). Instead, it focuses on the exact descriptions and accounts provided by research participants and the meaning they attribute to them in the context of a discussed phenomenon (e.g., a new technology). Exploring the rich narratives of the participants also enables the tracing of their everyday morality and the determination of how they arrive at potential or existing normative issues in relation to the technology in question, or how people define the normatively salient features of a technology in relation to their lives.

With this, the IPA method allows the identification of how the participants make a technology morally significant in their conversations rather than the pursuit of a moral reflection on the act of speaking or a conversation itself, a domain of the CA&DP method. Although the latter can also inform sense-making activity to a certain extent, the goal of the appropriation study is to trace and analyze the sense-making of people in relation to technologies, particularly the moral side of it. Therefore, for the purpose of my research, IPA can better aid in the investigation of technological appropriation than can the CA&DP method, as it can not only sketch the moral landscape of people but also identify how technologies mediate it.

In this regard, the shared philosophical roots of the technological mediation approach and IPA are very important. Both originate in the field of phenomenology, which focuses on exploring the lived experiences of particular individuals. From here, the shared adherence to micro-level studies as opposed to large-scale inquiries also originates. Particularly concerning the study of the human appropriation of technologies, hermeneutics takes center stage in exploring the interpretative sense-making of people. Similarly, both phenomenology and hermeneutics are central for IPA: "Without phenomenology, there would be nothing to interpret; without the hermeneutics, the phenomenon would not be seen" (Smith et al., 2009, p. 37). Thus, the shared philosophical assumptions behind the technological mediation approach and IPA suggest a fruitful relation between the two.

In view of these defining features of the CA&DP and the IPA methods, I must disagree with Verbeek's (2015) endorsement of the CA&DP as an empirical method to study appropriation, as the findings of the CA&DP would not fully satisfy the goal of the appropriation study and moral hermeneutics. Concerned with the sense-making activity of specific historically and culturally situated people, the content of and interpretations during conversation are of paramount importance for the hermeneutically oriented appropriation study. The IPA method seems to afford these dimensions of research. However, the specific fitness of the IPA method for the demands of the appropriation study also needs to be scrutinized in more detail, as does foreshadowing what it takes to make it fit with the philosophical analysis that the study of moral hermeneutics requires.

MAKING INTERPRETATIVE PHENOMENOLOGICAL ANALYSIS INTO A METHOD FOR STUDYING MORAL HERMENEUTICS

The IPA method can help uncover the relational experiences, both projective and practice-based, of people with specific technologies. As suggested above, unlike the CA&DP method, IPA can do so by focusing not only on how people express themselves but also on what they say, and crucially, by doing so not only descriptively but also interpretatively. Regarding moral hermeneutics, it seems that the IPA method allows capturing value dynamism through studying different parts of the appropriation process, as it can focus on the human sense-making while paying attention to the active role the material setting plays in it. Unlike the CA&DP method, IPA allows to capture the content of meaning attribution or the projective and practical interaction with a certain technology. With this, it can focus on the specific valuations and moral sensibilities that arise during the construction of meaning and trace how they are connected with the technology under study.

Crucially, with IPA, I will attempt to make a philosophical point supported by empirical and philosophical analysis. Using the conceptual framework of technological mediation as expanded by moral hermeneutics, I want to process and analyze empirical findings with the help of IPA. I wish to thus not merely describe and observe certain moral trends but constantly accompany IPA findings with philosophical analysis. This is the vision of empirical philosophy that I have outlined at the beginning of this book, and it is the one I want the IPA method to help me fulfill. However, since IPA originates in the field of psychology, I must also clarify how I intend to use it as a method for empirical philosophy.

A few defining features of IPA must be explicitly correlated with the exigencies of the study of moral hermeneutics through technological appropriation. The first concerns the field and scope of application of IPA. The IPA method originated in psychology, where it is still frequently utilized (Smith et al., 2009). However, its application extends into other fields, as does the use of the CA&DP method. While the experience of illness remains the top subject area of IPA studies, the scope of IPA research is wide, ranging from psychological distress to the experiences of education, music, alcohol, and information technologies (Smith 2011a,b). Commenting on transposing phenomenological research into the field of psychology (particularly with IPA), Smith, Flowers, and Larkin (2009) argue that "While philosophy has made an enormous contribution to understanding the process of examining experience, it is important to realize that philosophy does not own phenomenology" (p. 32). Paraphrasing the authors in pursuit of the method for sense-making with emerging technologies, one can equally say that psychology does not own IPA. In view of the above, extending the field of IPA research to the philosophy of technology and focusing on making sense of emerging technologies is not only possible but also falls within the broad research agenda of IPA.

The second IPA feature necessary to examine in relation to an appropriation study is the focus of the IPA study. The IPA method is concerned with studying the lived experiences of participants and how the participants themselves make sense of their experiences. The focus is on experiences that carry some existential import to participants, which are often related to certain normative issues that arise in the process. Connecting the IPA method with an appropriation study would shift the focus to new and emerging technologies, which would differ from the traditional IPA studies. Because such technologies either remain in the innovation pipeline or have only taken a first step into the market and are limited in quantity with restricted access, participants would generally not have direct experiences with such technologies. However, this does not preclude the use of IPA for an appropriation study, as IPA equally aims to scrutinize the "*perceptions* and *views* of participants (as alternatives to 'understandings')" (Smith, Flowers & Larkin, 2009, p. 46, original emphasis). This means that although direct experience with a discussed phenomenon is desirable, its lack does not preclude conducting an IPA study.

As the preliminary appropriation study of Glass in chapter 3 suggested, indirect experience with new and emerging technologies often precedes a direct encounter with a new technology. New technologies enter the minds of people early on in the form of news, public debates, anticipated benefits, hopes, and fears, as well as through advertisements, a company's reputation, and the experiences of early adopters with the development version of a new technology. In combination, this shapes the productive foreground of sense-making in relation to the technology in question and allows the exercise of proactive agency in pondering its relation to the participant's lives. Therefore, an IPA study can still be effective in discerning the sense-making of participants in relation to new and emerging technologies because it would focus on their perceptions and life views as well as invite them to exercise proactive agency as to how the new technology could relate to their historically and socially situated lives.

What this means for an appropriation study using the IPA method is the increased sensibility of the researcher to conducting interviews in such a manner as to engage the accumulated experiences of the participants and invite their projection onto the technology in mind. This requires, first, identifying and touching upon the topics of existential import throughout the interview and, second, relating them to the technology being studied and its capabilities. Combined with a thorough IPA analysis, this can comprise a rewarding process to identify how research participants make a certain technology morally significant in their life stories and how it can mediate—or already mediates—their normative conceptions.

Thus, even though the IPA method did not originally focus on the role of technologies in human lives, it can be put to this end without stretching the boundaries of IPA research too far. It can help the appropriation study to understand how people take up new technologies in their conversations and in reasoning about their lives and how the technologies in question relate to them. In doing so, IPA can help to reveal not only how people make technologies normatively significant in their conversations but also how they integrate them into their interpretative schemas, assimilating the understanding of a new technology with their existing preconceptions or rejecting them due to perceived incompatibilities.

Notwithstanding the suggested fitness of the IPA method for the appropriation study, it must necessarily be adapted to the cases at hand in some way while still complying with the rigorous analytical steps this method suggests. To identify and analyze the mediating role of technologies in the normative frameworks of people, in the following part of this chapter, I examine the specific technological case of the technology still in the making, the sex selection chip. Here, I will apply the IPA method accompanied by mediation analysis to try identify and explore the moral hermeneutics it might involve. The sex selection chip, contrary to Google Glass, COVID-19 tracing apps, or Ring doorbell, exists only in the prototype phase and in the form of newspaper and research articles, as it is facing stringent legal regulations in the medical and non-medical markets. Therefore, applying the IPA method to this case will put it to the fitness test in case of moral hermeneutics regarding an emerging technology.

APPROPRIATION STUDY OF TECHNOLOGY-IN-THE-MAKING: SEX SELECTION AND THE MORAL HERMENEUTICS IN CASE OF EMERGING TECHNOLOGIES

The potential to select the sex of a future child has been available since the early 1990s. Ethical concerns accompanied sex selection technology early on, ranging from worldwide population imbalances and gender discrimination to the fear of reinstating eugenic movements. Consequently, this technology faces strict regulation globally; it is only allowed when medically justified to prevent the transfer of genetic diseases or if national law endorses family balancing as a non-medical motivation. Currently, sex selection methods are expensive, invasive, and often require several attempts. However, a recently proposed upgrade of the technology in the Netherlands, a microfluidic chip for sperm analysis, suggests that a cheap, non-invasive method of sex selection in the setting of one's home may also be possible (MESA+, 2017; Valkenberg, 2014). In what follows, I will be referring to it as SST+, or an upgraded version of the sex selection technology (SST).

In contrast with Google Glass, SST+ is a new and updated version of a previously existing technology with established practices and strict legal regulations. Due to its controversial nature, the possibility of sex selection has early on sparked vibrant ethical debates worldwide. However, the new material setup and expected affordability of SST+ could enable new sex selection practices and challenge existing ones. In anticipation of this new technology and the change in the material context it suggests, an IPA study of SST+ could revisit the case of sex selection for non-medical reasons and explore the moral hermeneutics it entails.

To gain plausibility and avoid speculation in the absence of use experience, an appropriation study in the case of SST+ that anticipates its moral significance would need to rely more on literature analysis than in the case of Google Glass. In practice, this means that prior to conducting an IPA-based study, I must correlate the new technology against existing sex selection practices. This would imply exploring the technological setup of the sex selection practices and the ethical debate that surrounds them. Such a background study will allow me to understand what new concerns could arise in relation to SST+ and how these existing concerns could be reconfigured.

The application of the Interpretative Phenomenological Analysis method in this section aims to explore the potential of SST+ to mediate morality or to set moral hermeneutics in motion. This study will attempt to do this by identifying and critically reflecting on the anticipated manners in which SST+ could mediate the daily routines, practices, and perceptions of people in the Netherlands, the country where SST+ was developed. Note that passing a judgment on whether it is ethically justified to use SST+ is not the aim of the study. Rather, an appropriation study in this section, empirically oriented and accompanied by a technological mediation analysis, would aim to present a new angle in the sex selection discussion that is sensitive to the material changes that SST+ suggests and the everyday experiences of real people. To this end, I will first provide a brief technological and ethical analysis of the recently proposed upgrade to the sex selection technology, followed by an outline of the IPA study design and limitations. Then, I will present a snapshot of the IPA findings and conclude with reflections on the fitness of the IPA method for a study of moral hermeneutics through technological appropriation.

Technological Background

Assisted reproduction in the form of sex selection became available in the early 1990s. Currently, sex selection is possible via sperm sorting and pre-implantation genetic diagnosis (PGD) (Parliamentary Office of Science and Technology, 2016). In sperm sorting, two dominant methods are available: MicroSort, whereby distinguishing *in vitro* X and Y chromosomes is possible with the addition of fluorescent dye, and the Ericsson method, which relies on the higher mass factor of X chromosomes to sort the sperm when it passes through a protein like serum albumin. Conducting PGD requires the extraction of female oocytes for an *in -vitro* fertilization (IVF) procedure. Once the embryo develops eight cells, one cell is removed for chromosomal DNA analysis, which can reveal the sex of the embryo, among other characteristics (Harper & SenGupta, 2012). Because it provides higher result certainty than sperm sorting (99 percent vs. 75–85 percent), PGD dominates the market of sex selection (GenderSelect, 2017). However, because PGD necessitates the accompaniment of IVF, a successful pregnancy depends on additional factors and often requires several IVF attempts. The overall cost of sex selection is high, between 1,300–3,400 USD per attempt for sperm sorting, and 18,000–25,000 USD per attempt via PGD or IVF (ibid.).

Sex selection faces strict regulation worldwide. However, the prevention of the vertical transfer of genetic disorders such as hemophilia, Lesch-Nyhan syndrome, Duchenne-Becker muscular dystrophy, and Hunter syndrome justifies the medical use of sex selection (World Health Organization, 2011).

A non-medical exception concerns the desired variety of sexes in the family, when parents already have two or more children of one sex and want to have a child of another sex. Such a "family balancing" application of technology is available in the US, Cyprus, Ukraine, Israel, and several other countries (Bayefsky & Jennings, 2015, pp. 54–55). The technology behind sex selection is invasive (particularly with PGD and IVF), expensive, and requires multiple attempts. Despite this, sex selection for non-medical reasons is a top motivation for cross-border medical tourism (Van Hoof, Pennings & De Sutter, 2015) and accounts for up to 9 percent of the PGD and IVF cycles in the U.S., as of 2005 (Baruch, Kaufman & Hudson, 2008).

One recent technological development could significantly upgrade the existing sex selection methods and impact the practice worldwide (MESA+, 2017; Valkenberg, 2014). A microfluidic chip-based technology can measure sperm characteristics to ameliorate problems with the selection of spermatozoa common to assisted reproductive technologies. However, an adapted version of the technology could also be used for sex selection since the microfluidic chip can also analyze the chromosomal content of individual sperm cells (De Wagenaar et al., 2015, p. 1295). It would follow similar logic as the existing sperm sorting techniques: X chromosomes are longer and therefore heavier than Y chromosomes. The microfluidic chip operates at a nanolevel, where the chromosomal variation in weight would be significant for determining chromosomal sex, thus making sex selection in principle possible. One would need to present a sperm sample on a chip, which would sort the sperm into X- and Y-bearing groups, providing two options depending on the desired outcome. A chip-based form of sex selection (i.e., SST+), in view of its non-invasive nature, offers a home setting as a potential application site. Form-wise, it would likely take after its predecessor, as a point-of-care device, an at-home microfluidic chip for the assessment of semen quality, whereby a disposable chip, requiring small volumes of sperm sample, would be "used in combination with a handheld measurement system and management software" (Segerink et al., 2012, p. 66). Recent developments in microfluidic chips for sperm sorting even offer their use at home in combination with a smartphone for fast and cheap semen analysis and sorting (Khodamoradi et al., 2021). An at-home application would significantly reduce the overall costs of sex selection and expand its use beyond the medicalized domain.

It is critical to emphasize that the confidentiality of the research and its still predominantly open developmental avenues permit only a limited view of the technical specifications of SST+. Although researchers have confirmed the successful proof of concept of certain sperm characteristics (De Wagenaar et al., 2015), there is no word on pursuing the sex selection trajectory for human use. It is not clear how SST+ will manage the cells with an extra or missing chromosome (a chromosomal aneuploidy), although researchers previously used on-chip analysis that utilized staining protocols to successfully test for chromosomal anomalies in individual sperm cells (ibid., pp. 1299–1300). On a practical level, conducting the entire sex selection process at home implies not only the point-of-care sperm sorting promised by SST+ but also self-insemination, akin to using a turkey baster or a disposable syringe for at-home insemination using donated sperm. This approach presents limitations and often necessitates multiple attempts at pregnancy (Wikler & Wikler, 1991). Although the form and usability of the technology remain unknown, anticipations of a chip "for 12.95 at the drugstore" (Valkenberg, 2014, ¶16) suggest an expected resemblance to a pregnancy test in terms of cost, ease of use, and acceptability. Some anticipate that "The [sex] selection chip [. . .] has the potential to become available at a large scale and a low price; therefore, the social effect is likely to be quite substantial if the technology would indeed be introduced" (Verbeek, 2015, p. 197).

Importantly, as mentioned earlier, sex selection for non-medical reasons remains illegal in many countries, with rare exceptions for family balancing. At the same time, history suggests a close intertwinement of technological innovation, public views, moral frameworks, and legislation. Initially, sex selection appeared to increase the birth of female animals to boost agricultural outputs, but it later migrated to use in humans. Both the research and the history of sex selection identify no technical obstacles to using SST+ for people and indicate that it is in principle possible (Valkenberg, 2014).

In what follows, I wish to consider the potential leap of SST+ to human use, although I want to stress that no support for pursuing this agenda exists among the technology's developers. Nonetheless, such an anticipative moral hermeneutics study is useful and necessary as a test bed for considering the previously unacknowledged moral significance of the material setting in the sex selection practice. Moreover, in the hypothetical event of market introduction, this study can offer matters of concern for the responsible design and governance of SST+, a technology that still has an open future regarding its form, usability, and societal embedding.

Ethical Debate

Ethical considerations regarding the societal risks of sex selection are what have ultimately informed its strict regulation for non-medical reasons. The debate around original sex selection technology encompasses diverse ethical concerns and anticipated benefits. The proponents of this technology rely on rights-based claims to suggest that people should be free to make reproductive choices without guidelines from the government or anyone else, insofar as these choices do not limit the freedoms of others (Dickens, 2002; Savulescu, 1999). The case against sex selection is mirrored in the worries of catering to the preferences of prospective parents and disregarding the inherent worth of a child (Sandel, 2004), as well as suggesting that this technology is discriminative by design, promotes sexism, and will foster unreasonable gender expectations in prospective parents (Blyth, Frith & Crawshaw, 2008). Contrary to the traditional sex selection methods, the use of SST+ suggests a certain and safe selection outcome at a low cost and with home use, which would also eliminate the need for embryo creation. While this change in the material setting of the sex selection practice may address some of the ethical concerns outlined above, it does not tackle all of them. The question is rather whether SST+ could affect society in ethically nuanced, subtle ways that could gradually inflict social and normative change in both individual and public views. In other words, whether SST+ could mediate the moral landscape of human beliefs and practices.

Having conducted a thorough literature analysis, I outlined several directions in which SST+ with its new material setting could impact the current ethical debate surrounding sex selection (Kudina, 2019). Namely, I suggested that it could enable a new set of sex selection practices, fundamentally change their nature, foster new manners of perceiving sex selection from an individual and societal point of view, and crystalize new power relations in the process of childbearing.

Firstly, the move of sex selection to the setting of one's home could demedicalize and normalize the practice of sex selection. With this, SST+ could reinforce the values of naturalness and privacy for those who seek this procedure. In parallel, SST+ could highlight the role of men in the reproductive process, which could mediate parental relations in multiple ways, ranging from leveling the power balance in childbearing to abusing newly acquired patriarchal authority.

Secondly, making SST+ an affordable direct-to-consumer technology could shift the self-perception of its potential users from patients to customers, entitled to the right of choice, and from parents abiding fate to parents actively determining it. Such shifts could co-shape new normative understandings of what good parenthood means. In parallel, significantly simplifying the process of sex selection and reducing its price could foster a societal "Why not?" attitude, making private reproductive choices available for potential justification.

Thirdly, the potential move of SST+ to an open market could mediate the value of responsibility in relation to the various stakeholders involved in its introduction and entry to the market. The introduction of SST+ could foster different manifestations of parental responsibility in the course of evolving personal histories and sociocultural contexts; contextualize the industry and marketing responsibilities toward ethical reflection beyond immediate technological use; and maintain the roles and responsibilities of medical professionals in identifying the myriad nuances that could materialize during the use of SST+.

These preliminary findings point to the moral hermeneutics potential of SST+ identified at the level of literature studies and insights into current technological practices. To further substantiate the moral hermeneutics potential of SST+, requires attending to the micro-perspective of the real-life experiences and concerns of individual persons. The following sub-section builds on the literature findings identified here and presents an IPA study of how people could appropriate SST+, as well as the ensuing value mediations.

The Setup of the IPA Study

This IPA study aims to understand how Dutch citizens appropriate SST+ for non-medical reasons: how people reason with it, attribute meaning to it and position it in their lifeworld, and normative frameworks. I have focused on citizens of the Netherlands for several reasons. Firstly, SST+ is a Dutch innovation, and hence it is important to understand the position of the people who might be affected by its introduction, both directly and indirectly. Secondly, a study of public opinion in the Netherlands regarding the early versions of sex selection (i.e., the Ericsson method) and its use for non-medical reasons already exists, commissioned by the Rathenau Institute (1996). The Rathenau study indicated that although Dutch citizens did not have specific preferences regarding whether to have a girl or a boy, 78 percent of respondents condemned the use of this technology (Rathenau Institute, 1996; Volkskrant, 1996). However, the study occurred more than 20 years ago and focused on surveying large population groups, differing in the nature of the study and the technology in question from the current investigation. As such, the present study approaches members of the Dutch population to understand, from the perspective of their situated life experiences and contexts, how they make sense of the new sex selection chip and the possibility of its use for non-medical reasons in the Netherlands.

"Dutch population," just as any general cumulative concept, is an abstract phenomenon that requires construction and definition. Guided by the IPA methodology described earlier in this chapter, the study does not strive for statistical representation and targets a small number of participants to study them in detail for a close, rich picture of their modes of reasoning. This IPA study regarding the appropriation of SST+ consisted of seven diverse participants, bound by a group belonging to Dutch culture. To ensure the richness of perspectives, it was important to include people from different backgrounds, professions, and geographical areas, provided they were all adults who were born and resided in the Netherlands. The resulting heterogeneous group of participants represents a wide range of educational and professional backgrounds, including students; employees in the areas of catering and entertainment; administrative and clerical staff; social workers; PhD candidates; and those unemployed at the time of interview. The group spanned the ages of twenty-six to fifty-six, with four men and three women. Because people inhabiting the central area of the Netherlands known as Randstad seem less hesitant toward technological innovation, qualitative studies of the Dutch population and new technologies suggest including people from different provinces in the Netherlands to ensure a plurality of perspectives (Schuijff & Munnichs, 2012, p. 53). To decrease this potential bias, study participants were recruited from different parts of the Netherlands: one from Limburg, two from Noord Brabant, two from Gelderland, and two from Overijssel, with varying sizes of cities. Participants in the current IPA study thus did not form a homogeneous group in view of age or professional background, but their cultural and geographical belonging to the Netherlands united them.

The study in this section examines how people make sense of a technology that is still in the innovation pipeline. Therefore, the participants did not possess first-hand experience with it and rather relied on understanding the principle behind SST+, with minimal information prompts from me. In this way, I, as a researcher, fostered the proactive agency of the participants to form their opinions and exercise their judgments regarding the technology-in-the-making. Their lived experiences were equally involved, with the interview questions targeting their experiences of being a child and their relation to childbearing and parenthood, as well as living in the Netherlands and considering the expected impact of SST+ on living together in Dutch society. The current IPA study has thus attempted to obtain an informed picture of the technological appropriation process based on both proactive agency and the lived experiences of the participants.

Before the interviews, I contacted the study participants to obtain their informed consent for the study and received permission from the Ethics Committee at the University of Twente to ensure that the study corresponded with ethical standards. Depending on the participants' preferences, I interviewed them in Dutch or English. I audio recorded the interviews and manually transcribed them with MS Word. I withheld the identities of the participants and gave each of them a pseudonym, distinguished in the text with italics, e.g., *Matthias*. The analysis of the data followed the IPA methodology steps outlined earlier in this chapter. I additionally used color-coding and excerpt numbering techniques to distinguish between different concerns raised within the interviews and develop emergent themes across the interviews.

In an attempt to distill and reflect upon the specific value constellations of the interviewed Dutch people, my cultural belonging to a different geographical region unexpectedly aided me in this task[2]. Although this inevitably projected other cultural hermeneutic layers from my interpretative structures, it also provided me with a sense of a fresh perspective, being both an insider and an outsider in the process of analysis. With this in mind and following the IPA method, I identified five overarching superordinate themes, with each appearing in at least four out of seven interviews and representing a particular concern related to SST+, or following the idiosyncratic commitment of IPA, when the theme was dominant in only one or two interviews but nonetheless raised an important issue for the study.

The IPA Findings

The superordinate themes identified in this study each present a unique normative concern as well as its malleability when faced with SST+. The identified themes concern (1) understanding good parenthood; (2) relating SST+, gender, and culture; (3) enforcing the trend toward perfectionism; (4) mediating naturalness; and (5) considering the value of liberalism and choice. All of the themes consider SST+ in relation to certain values, existential concerns, or normative contexts and, as such, represent the anticipated potential of this technology to spur moral hermeneutic processes in motion. In what follows, for reasons of space, I give a brief snapshot of the first four themes and present the fifth one in a detailed write-up analysis.

The first theme concerned *understanding good parenthood in the face of SST+*. It was the most dominant theme running throughout all interviews and reflected similar concerns from all seven participants. Overall, the rich canvas of beliefs, values, and opinions that the participants presented brought to the fore the mediating force of SST+ in relation to the values of parenthood. On the one hand, SST+ appears as a conditioning technology, confronting the values of unconditional love, openness, and acceptance that the participants see as pivotal for parents, or as witnessed by an interview quote from the participant *Lucy*: "You have a child, and then you have to do with whatever you get." On the other hand, SST+ appears in the participants' narratives as a materialized link between sex and gender and the material manifestation of their gender preferences. The participants highlight the apparent connection in the eyes of the public between sex and gender, a complicated relationship that is easy to conflate. By presenting a choice between two options, SST+ might foster this link between the two concepts in the minds of parents. The participants consider fostering gender expectations in the children of the selected sex a danger, with possible complications of parent-child relations, the active projection of gender expectations upon children and the potential for psychological pressure on them.

However, SST+ need not result in pressure on children if parents uphold the value of flexibility and openness to the agency of children in identity building and decision-making. As such, the participants identify the mediating effects of SST+ on parental values and parent-child relations and touch upon the complicated sex-gender relations in the context of family building.

The second theme that resurfaced in six interviews out of seven more explicitly focused on *relating SST+, gender, and culture*. Many respondents remarked upon the difficulty of separating the concept of sex from the concept of gender, oftentimes equating the two. It is against this background that many respondents positioned SST+ to understand how it fits the domains of family and society that they inhabit. The IPA study suggests that SST+ could further complicate the already complex and precarious relationship between sex and gender in Dutch society. SST+ could potentially suit the needs of those seeking procreative freedoms but complicate the identification process for those struggling to fit traditional sex and gender categories in view of dualistic gender expectations. For these reasons, the participants described the technology as deeply problematic and, as quoted from *Hendriks*, designated SST+ as "A step in a wrong direction."

The third IPA theme mirrored *the worries of SST+ enforcing the trend and pressure toward perfectionism*: "It's like you have to sketch your entire life" (*Melanie*). It identifies a close connection in the eyes of the respondents between SST+ and a perceived societal trend toward perfection. The theme of perfection appears differently among respondents, for instance, in the family layout, in a desire to have children of both sexes, and on a societal level, represented by an idea of social status. The respondents were critical of the idea of perfection and associated the potential introduction of SST+ with the pressure to fit into the "perfect society," where many choose to select the sex of their child. Their responses also agree regarding how SST+ appears as a technology both promoting perfection and fitting into the perceived societal trend, something that they ultimately consider undesirable.

The fourth theme scrutinized *the value of naturalness as mediated by the mere idea of SST+*. The two ways in which this theme surfaced concerned the value of naturalness in pregnancy and the value of naturalness in itself as a driving force in life. Even though the respondents acknowledged the present big role of technologies in mediating what "being natural" means, from the color and shape of tomatoes to the *In Vitro* Fertilization technologies, overall they were skeptical regarding the way SST+ can meddle into a gift or a miracle of life that SST+ could only threaten or damage, as exemplified by a quote from *Anouk*: "May we really determine everything?"

Finally, the fifth dominant theme from exploring the potential appropriation of SST+ concerned its *relation to the values of liberalism and choice*. Here, unlike the zoomed out reflections on the themes above, I present a detailed excerpt of the IPA study, accompanied by the extensive participants' quotes and the technological mediation analysis. This theme reflects how the respondents consider the possibility of sex selection for non-medical reasons in relation to individual freedoms and rights, particularly in the context of Dutch society, which is described by most participants as liberal and respectful of individual choices. In at least four interviews, the discussion of liberalism was explicit. Three respondents reasoned with SST+ in direct relation with euthanasia and assisted dying in general.

The excerpt of the interview with *Matthias* below presents a rich, value-laden narrative of attempting to make sense of SST+ in view of traditional norms and current developments in the Netherlands.

Matthias: At the moment it could be that if it [allowing sex selection SST for social reasons] were put in the Parliament, I'm not really sure, I think it has a chance to be more legal, actually.
Olya: Why is that?
Matthias: Because confessional parties are less strong at the moment, the Christian parties. So I think the liberal parties are more in favor or accepting it as a choice. . . . The discussion in Holland is, on the other side, about euthanasia. That's more of a discussion, should you be able to say at a certain age, to say I've had enough. And even there parties are in favor for, almost in majority in parliament. So to make it even more liberal as we already have. So then I think, if you're that easy about death, then to make a choice like this… Although it's a different subject, it has different merit. In the other case, we think of a person that's already there and can decide for themselves. And of course with this technology it's about someone who does not exist yet, so it cannot decide for itself, to make certain decisions about its own life. But I still think there is. . . . I think there is maybe a move towards more acceptance than in the past. . . . The move is to have more options even when you're not ill and more freedom for people to make decisions. I think you should be really careful. . . . And also the question should be, should a state provide these options. It's also a thing. I would prefer a state that does everything to keep me alive [laughs], more or less.

In this excerpt, the possibility of legalizing sex selection for non-medical reasons fosters *Matthias* to reflect on the value of liberalism in Dutch society, in the face of the expanding technological options. *Matthias*, concurrent with the literature findings in the previous subsection, frames the possibility of sex selection as a choice and, on the political level, as the right to choose. He considers the Netherlands a liberal society, one that provides and respects the right to choose. Although *Matthias* supports the right to choose, he appears hesitant to embrace all possible (technological) options that further expand liberalism, for instance, in the case of sex selection for non-medical reasons. One can intuit this hesitation from how *Matthias* cautiously compares SST+ with the present-day political discussion in the Netherlands on assisted dying ("euthanasia"), particularly regarding expanding its legality for non-medical reasons. Amid this comparison emerges a critical reflection on the value of liberalism and on how certain opportunities, albeit possible, might not be desirable overall.

From the words of *Matthias*, one can interpret liberalism as a value of free choice and decision-making regarding matters concerning one's own life. He provides an example of the current political discussion in the Netherlands about legalizing the right "To say I've had enough," or assisted dying (euthanasia and physician-assisted suicide) for anyone tired of living, even without medical reasons causing unbearable suffering (Uit Vrije Wil, 2017). According to *Matthias*, although Dutch society in principle favors individual freedoms and choices, he currently observes a trend toward further expanding individual liberties: "The move is to have more options even when you're not ill and more freedom for people to make decisions." However, *Matthias* believes that assisted dying and SST+ fundamentally differ regarding agency. It appears that SST+ would challenge the value of liberalism by providing a choice about the future for not yet existing people who cannot make the choice themselves. In this manner, the liberal approach to legalizing assisted dying, where people can decide on their own lives, would not fit the case of SST+ for non-medical reasons. This, in parallel, sketches the limits of individual choice and the right to choose in the case of SST+. Ultimately, *Matthias* appears to question to what extent the value of liberalism actually concerns SST+, and he cautiously suggests that liberal pro-choice policies need not always apply.

The subject of limits to individual liberties in a liberal society also resurfaces in an interview with *Hendriks*:

Olya: What do you think about this possibility in general [SST+ for social reasons]?
Hendriks: It's really hard. I have several points, you know. Maybe one point, I really like this idea that a person can decide for himself what everyone wants, in case of abortion, in case of life ending. . .Yeah, it's really important for me that there's no society saying that no, you can't abort or no, you can't ehm . . . your life . . . end it yourself. I really like this personal ehmm . . . ability to choose whether or not. But I think with this sex selection, you're actually not choosing for yourself but for the person that is about to come to the world. That makes societal opinion about it a little bit more legitimate. You know what I mean? [. . .] I think the idea with liberalism is not that you would be free in A-A-A-LL aspects. [. . .]
Olya: What is the relation between sex selection technology and liberalism?
Hendriks: [pause] Maybe that it is an aspect of a very liberal society that you as a society would not want. . . . There are issues at hand when even . . . even in a liberal society you would not want to fit together.

For *Hendriks*, as for *Matthias*, the idea of liberalism, "that a person can decide for himself" in a society that does not restrict individual liberties, is of existential importance. He approvingly cites cases of the rights to abortion and assisted dying as examples of such liberal rights. However, on par with *Matthias*, he draws a stark contrast with sex selection and suggests that liberalism has its limits: "The idea with liberalism is not that you would be free in A-A-A-LL aspects." Because in the case of sex selection, one chooses not for oneself but for another person who is yet to enter the world, sex selection warrants societal concern and distance from the liberal principles of respect for one's choices. He questions the desirability of SST+ on a societal level and generalizes his critical approach to the technology by suggesting that "it is an aspect of a very liberal society that you as a society would not want." With this, *Hendriks* suggests that SST+ does not contradict the value of liberalism; rather, it highlights one aspect of it—the right to say no to what is offered.

Paul, considering the potential future of SST+ in the Netherlands, also distinguishes between the right to choose and the right to say no:

Paul: I don't have problems with it [SST+]. I would not do it but I would not have problems with it. Some people, they think it's important. Yeah? And what's important for them—well, you have to give them a chance. . . . But I would not choose for it. But the same, if you're older and you think life is not responsible to live—and it must be also possible. You understand what I meant?
Olya: Maybe not, can you explain a bit more?
Paul: Some people they are old, and they are not in good condition. And they want to say, I want to end my life. And I'm also saying, that must be possible. Thus, this one must be possible also.

In this excerpt, the value of liberalism emerges through *Paul*'s considerations of allowing people to do what they find important and retaining the possibility to refrain from a certain action (as he himself expects to refrain from SST+). *Paul* presents the core idea of liberalism, when he suggests that if people find something important, they must have the chance to fulfill it. Similar to *Matthias* and *Hendriks*, *Paul* draws a parallel between ending one's own life and SST+, but unlike the other respondents, *Paul* does not distinguish between assisted dying and SST+, and he applies a general principle: if one is permitted, then the other must also be permitted. As such, SST+, for *Paul*, does not challenge the principle of liberalism but rather falls under its wide umbrella: because both assisted dying and SST+ represent matters of concern and importance for people, in both cases, people must have the right to do what they consider best.

In contrast to the other respondents above, *Lucy* places liberalism in the context of parental virtues to draw the line on individual freedoms:

I'm finding it difficult really to say that people . . . if they really want it, shouldn't be able to use it. But in the end I don't think it's emm . . . it's a good thing. No . . . I think it's good to have as little choice about what you're getting as a child as possible. For *Lucy*, SST+ fosters a value conflict between her appreciation of liberalism and her understanding of virtuous parenthood. On the one hand, she struggles to deny people the use of sex selection if they express such a desire. On the other hand, as we have seen above, a good parent, in the eyes of *Lucy,* possesses the virtues of acceptance and unconditional love. Upon consideration, *Lucy* finds that the values of good parenthood trump the value of liberalism in making the choice to select the sex of one's future child: "It's good to have as little choice about what you're getting as a child as possible."

Overall, the value of liberalism in relation to SST+ appeared in discussions on individual choices and freedoms and their scope and applicability in a society that the respondents define as liberal. Three out of four respondents struggle to define the limits of liberalism in a society that traditionally respects individual freedom of choice, but they still suggest that SST+ could be an example of such a restriction. Particularly for one respondent, in the case of SST+, the parental virtues of embracing the uncertainty that a child brings appear more important than the parental right to decide on their family. Another respondent considers that if using SST+ is important to people, they should be granted such an opportunity, given that Dutch society permits other life-related decisions such as the right to end one's life with age. However, particularly in comparing SST+ with assisted dying, other respondents suggest that liberal principles of choice need not apply to SST+ because the decision-making concerns people who are not yet conceived and cannot make decisions about their own lives. According to these respondents, approaching SST+ with the value of liberalism in mind contradicts its constituent value, namely, agency in the matters of one's own life. Overall, the value of liberalism comes to the fore as dear to most of the respondents, who define and re-articulate its scope in view of the possible implementation of SST+.

Summing up the first attempt at the IPA study, one can see how this method, applied to the case of SST+ for non-medical reasons, has generated a rich account of how participants appropriated SST+ (i.e., positioned and fit it in their frameworks of understanding). This method provided a nuanced methodology to interpret and reflect on how projective SST+ appropriation varied among the participants, generating overarching similarities while maintaining contextual sensitivity. Although none of the participants possessed first-hand experience with SST+, using in-depth interview questions as prompts and relying on their own experiences and proactive agency spurred participants to display an understanding of the technology and a certain attitude toward it.

While most of the participants would not consider the introduction of SST+ desirable, their motivations for such a position diverged.

Some of these themes suggest a potential value conflict that the possible introduction of SST+ would pose for the moral landscape of the participants. For instance, considering good parenthood, the majority of respondents suggested that SST+ could violate the values of acceptance, unconditional love, and dealing with uncertainty that the respondents identified as cornerstones of good parenthood. For some respondents, SST+ would further undermine the value of naturalness in letting life take its course or represent the manipulation of nature, something that certain respondents considered inherently wrong. Considering SST+ in relation to pregnancy enabled some of the respondents to actively reflect on what they consider important, not only to themselves personally but also with respect to societal values. As such, one of the respondents identified the overarching, enduring Dutch culture of naturalness in all pregnancy-related practices, where using minimal technological intervention is valued. In this regard, SST+ would reaffirm the value of naturalness in conception, since the respondent regarded the possibility of sex selection as an insufficient incentive to give it up. Overall, presented with the possibility of SST+ in the Netherlands, the respondents revealed what is important to them and how it is mediated by the mere possibility of SST+.

Some themes indicate the negotiation of certain values and an active reflection on and re-articulation of the concepts that elicit concern among the respondents. For instance, one of the most dominant themes in the study concerned the complicated relationship between sex and gender and the role of culture within it. Confronted with the possibility of SST+, the respondents identified several ways in which SST+, in their eyes, could fit this dynamic. Most respondents upheld the values of equality and liberalism that they associated with Dutch society. On the one hand, SST+ could fit the value of freedom of choice by allowing potential parents to determine their family layout. On the other hand, the respondents predominantly regarded SST+ as a value-laden technology, attributing it with the possibility of promoting dualistic gender expectations and stereotypes.

The respondents suggested that SST+ could make the difference between sexes explicit and available for choice. The fear here concerned reinstating the traditional societal layout that, according to the respondents, relied on dualistic gender identities, expectations, and stereotypes that directly linked sex and gender. From this angle, SST+ presents the need to reinforce the values that the respondents identified as shared in the Netherlands. For the majority of the respondents, this entailed considering the introduction of SST+ as undesirable in the Netherlands. This also outlined the limits to the value of liberalism, which most proclaimed as also defining Dutch society. However, one respondent insisted on making SST+ possible for those who desire such an option, suggesting that a liberal society should do what it can to enable people's happiness. Interestingly, those respondents who opposed the possible introduction of SST+ also reasoned with the value of well-being in mind. Certain respondents suggested that the possibility of one sex or the other, in the context of the complicated and often conflated relationship between sex and gender, could complicate the lives and, arguably, the well-being of people who do not identify with or fit into dualistic sex or gender categories. They suggested that the introduction of SST+ in Dutch society could be unproblematic only if a clear distinction between sex and gender and a clear understanding that SST+ could offer a choice of only certain biological features could exist.

The narratives of the respondents also indicate an openness to the possibility of moral change regarding SST+, illustrated by the following exemplary quote from *Melanie*:

> There are more and more things that we consider nowadays as normal, but previously we did not. Then I think, yes, maybe in 20 years, "Huh, have I ever said that?" Then it might be completely normal that you choose for a boy or a girl, and I'd think that.[3]

This respondent not only acknowledges the idea that values can change but also foresees this occurring in her lifetime. The change in values and moral views is presented here as a dynamic element of life.

Notable in the discussion on the mediating role of technologies in moral hermeneutics is an excerpt from the interview with *Hendriks*. Hendriks defines himself as coming from a conservative, religious farming family. Before we even began the interview, he described an encounter with sex selection in his family. The context was farming, where having cows was essential to generating farming products and, as such, responsible for earning any profit.

> *Hendriks:* My parents have a farm with cows and there they apply sex . . . not this specific technology but the main technology of selecting . . . eh like female cows. . . . My parents are very Christian. So at first, it's really funny, at first they were like, "Oh no-no-no, we will never do that! It's like . . . it's like messing up God's plan!" Then, I think it started a little bit with irritation. It's really irritating if you really could [do that but you don't]. For my father, for example, when he then for the second time, the THIRD time in a row gets a male calf, you, I think at a certain point, you're like, well, maybe we change for one [quietly] and then for two [laughing]. . .so slowly [it gets accepted].

This passage exceptionally clearly illustrates moral hermeneutics with the example of sex selection in animal farming. Sex selection technology here mediates the relation between the attained values of naturalness and religious guidance, on the one hand, and the value of profit (rather, the lack of profit) precisely in view of letting nature take its course, on the other. *Hendriks* reflects with irony on how technology slowly modifies initially held beliefs to achieve a new balance between all concerned values.

Overall, the IPA study unveils the different layers of moral hermeneutics—the mediation of different families of values and their underlying moral concerns—that resurface through the projective SST+ appropriation. The rich findings of the study visualize the active and unruly process of attempting to make sense of a new technology by referencing it with existing individual experiences and knowledge and, depending on this, revealing, at times unique, at times shared, value constellations and possible conflicts. The IPA study of SST+ foregrounds the interpretative appropriation process of the respondents, where the dynamic moral sense-making appears as an inalienable counterpart of technological appropriation. Although the respondents did not possess first-hand experience with the technology, their rich life experiences allowed for the anticipation of future use practices and specific scenarios regarding the introduction of SST+ in the Netherlands. As suggested by the IPA method, the proactive agency of the interviewees allowed them to engage in explicit and implicit technological reflections and valuations, which presented a rich foundation for an IPA study. In attempting to make sense of SST+ and position it in their lives, the respondents revealed their existential concerns, identified value conflicts, and revised certain conceptions to negotiate conflicting meanings while also reinforcing the understanding of others. In this manner, the IPA study provided insight into how respondents appropriated SST+ and visualized the study of moral hermeneutics and the dynamic process of morality-in-the-making.

CONCLUSIONS

This chapter sets out to develop an empirically philosophical method for a systematic study of moral hermeneutics. Such a method would need to be able to capture the dynamics of human interpretation both when people make sense of technologies that already exist and when the technologies in question are still at the threshold of innovation. Being able to capture both the practical and projective appropriation of technologies would be essential for demonstrating the moral sense-making that is a counterpart to technological appropriation. A method for moral hermeneutics must be able to demonstrate how, in an attempt to fit technology into their lives and interpretative schemas, people also bring to the surface usually tacit moral concerns and values that guide them through life, reviewing, updating, or making space for new values.

In view of this explicitly hermeneutic core, a method for the study of moral hermeneutics would also need to account for the subtleties of human interaction and experiences, be attuned to how people express themselves, and make themselves heard explicitly and implicitly, e.g., through evoking certain metaphors or changing the way they talk. Therefore, an empirical method for the study of moral hermeneutics would require attention to both the content and the form of human interaction. Because of these considerations and after a thorough methodological review, I opted for the Interpretative Phenomenological Analysis (Smith, Flowers & Larkin, 2009) method over the one of Conversation Analysis and Discourse Psychology (Edwards & Potter, 1992; Te Molder & Potter, 2005) as the most suitable method for the study of moral hermeneutics.

Throughout the chapter, I considered what it would take to adapt the originally psychological method of IPA to the philosophical study of moral sense-making through technological appropriation and tried its fitness in the case of an emerging biotechnology, the sex selection chip. Apart from the considerations of an active engagement of the researcher in the interpretation processes and the difficulty of integrating the parallel philosophical analysis to the rich empirical descriptions, I would also like to mention another exigency of developing and applying empirically philosophical methods: its highly demanding and time-consuming nature. As suggested by Boenink and Kudina (2020), "Uncovering the dynamics of valuing processes in relation to new technologies is labor-intensive and, because of this, often tends to focus on a limited set of cases, practices and/or situations" (p. 15). As can be seen from the sex selection IPA study, a thorough literature analysis may often be required as a pre-study for the IPA interviews. The interviews themselves are labor-intensive in that the hermeneutic and idiosyncratic nature of the IPA method favors manual approaches over those aided by software-aided analysis. Coupled with the micro-focus on several study participants, all these considerations may limit the broad appeal of the qualitative empirically philosophical methods to study morality-in-the-making, with IPA as one instance. Nonetheless, such limitations can also be taken as the strength of the method, owing to its ability to demonstrate the rich lived realities of the participants, evoke their proactive agency in the case of technologies still in the making, and make the entangled canvas of normative valuations available for reflection. Crucially, a method such as IPA can do so in a methodologically robust manner, making each step of the process available for bystander review.

Taking into account both its strengths and limitations, and after successfully applying it to the case of technology at the threshold of innovation, I can conclude that the IPA method is helpful for an empirical exploration of moral hermeneutics mediated by technologies. What remains in my overall aim for developing an encompassing understanding of moral hermeneutics is an elaboration of a philosophical principle of moral sense-making that would combine all of the findings from the earlier chapters. Namely, such an encompassing principle of moral sense-making must be able to incorporate the conceptualization of morality as a dynamic and evolving ecosystem, including at least humans, technologies, and the larger cultural setting; and the idea that values, following the pragmatist tradition, can change in interaction with the sociomaterial environment that both enables them and is guided by them. This is what the following chapter sets out to do.

NOTES

1. Idiographic research (e.g., Schwandt, 2007, p. 145) focuses on in-depth detailed studies of individual socially and historically situated cases, relying on the exact descriptions and experiences of the participant in trying to make sense of an event. Idiographic research methods include case studies, interviews and any other methods focused on holistic representation of the research participants. Idiographic research is often contrasted with nomothetic research, which focuses on large-scale research groups with a goal of generalization across broad populations (e.g., survey and questionnaire methods, quantitative research methods). Idiographic methods often belong to the domain of qualitative research, while nomothetic to the quantitative research.

2. The author is a Ukrainian living in the Netherlands.

3. Translated from Dutch. Original: "*Melanie*: Er is tegenwoordig ook steeds meer normaal wat niet normaal was. Dan denk ik nou ja misschien over 20 jaar Huh heb ik dat ooit gezegd? Dat het dan heel normaal is dat je kiest voor een jongen of een meisje, dan vind ik het wel, dan denk ik dat."

Chapter 5

Hermeneutic Lemniscate as an Encompassing Principle of Moral Sense-Making Mediated by Technologies

In this chapter, I synthesize the conceptual and empirical explorations presented earlier in this book to develop an encompassing principle of moral hermeneutics. To do so, I will use the concept of technological appropriation, presented in chapter 3, and the expanded understanding of the technological mediation approach, outlined in chapter 2, to explain the role of people, technologies, and the sociomaterial world in the joint co-production and reinterpretation of moral meaning. Throughout the chapter, I will draw on the case of voice assistants for illustration of the concepts and the urgency to develop an encompassing principle of moral sense-making. Digital voice assistants (DVAs) redefine the way we interact with technologies, presenting voice as the primary interface. Natural language processing algorithms and other forms of artificial intelligence underlying DVAs allow people to use phones for dictating messages rather than typing them or asking the phone to set an alarm. Increasingly, voice interfaces permeate the homes of people in the form of smart speakers, such as Amazon's Echo with Alexa as a virtual assistant or Google's Home with Google Assistant. But beyond turning on the lights and informing us how far the nearest pizza place is, DVAs also change the way we interact with each other.

As shown in chapter 3, technological appropriation is closely linked to the ethical implications of technologies because it inevitably embeds moral sense-making. Because values are an inalienable counterpart of our sociocultural and individual histories, attempting to understand a technology also engages normative intuitions, making them visible and available for renegotiation. Even though the first DVAs appeared only a few years ago, there are already signs of their social and ethical implications. Voice assistants process users' speech as commands or requests, discrediting the accompanying niceties, jokes, or sarcasm as non-functional statements. The users learn that if they want Siri or Alexa to understand them, they need to be as concise as possible. However, DVAs are not perfect at processing human speech, often asking to repeat the questions multiple times and unchallengingly accepting angry or offensive responses from the users. Most often, this is done in a female-sounding voice. As I will argue in this chapter, such design features of DVAs recontextualize the meaning of the world to the users, which in turn helps to shape them as specific subjects. This may change not only how we understand ourselves but also what we expect from those around us.

Technological appropriation can be a valuable entry point to explore moral hermeneutics by elucidating several matters related to technologies and meaning-making. Firstly, the sense-making activity underlying technological appropriation involves the interaction of (at least) three actors: people, with their epistemic structures and beliefs; technologies, representing a phenomenon that requires integration into the epistemic and practical frameworks; and the sociocultural world, as an active context against which a specific human-technology encounter occurs. Secondly, the three dynamic and interrelated elements of the appropriation process prevent it from being a static, final event. People can reconsider the meaning and place of a certain technology in their life once the existing meaning no longer fits the current situation. The stability of a preliminary meaning bestowed upon an artifact depends on the interaction between people and technology in a specific sociocultural setting. However, the appropriation remains an open process.

While the adoption of DVAs is still ongoing, I draw on this example to explore philosophical challenges regarding the inclusion of technologies in the sense-making processes. Even the preliminary examination of DVAs above shows that we cannot avoid (re)inventing ourselves while making sense of technologies. Technological appropriation thus appears to be a promising concept to further untangle the relation between technologies and moral hermeneutics. However, the philosophical principles underlying the appropriation process require more elaboration, as well as the mediated subject constitution that occurs not only during but also after technological appropriation. To theoretically substantiate the account of moral hermeneutics, I will now turn to the hermeneutics of Gadamer (1975/2004; 1977) and the material hermeneutics of Ihde (1990; 1998) to see how they account for the sense-making processes of people. Combining the insights, I will next offer a novel theoretical account of technologically mediated *hermeneutic lemniscate* that explains how people, technologies, and the sociocultural world jointly produce and maintain the meaning-making. I will illustrate the moral dimension of the hermeneutic lemniscate by exploring the case of DVAs to outline how they influence our perceptions, actions, and moral intuitions through specific technological mediations. In the final part of the chapter, I will suggest a few philosophical implications of the lemniscatic principle, as well as on the use and design of voice assistants, that will pave the way for the conclusion.

FINDING A PLACE FOR TECHNOLOGIES IN THE PROCESS OF SENSE-MAKING

Hermeneutics and Gadamer's Circular Account[1]

I would like to briefly explain the hermeneutical foundation of the nature of interpretation and suggest that it requires an expansion to address the challenge of considering the role of technologies. Gadamer's hermeneutics is particularly useful here because it explains the dynamic nature of interpretation, its historical imbedding, and the role of productive prejudices. Even though other hermeneutic scholars can also shed some light on the interpretative processes related to technologies,[2] Gadamer's work is particularly interesting in relation to the moral mediation of technologies. Gadamer (2004/1975) positions moral ideas and beliefs in the effective fore-structure of human understanding, where they are intertwined with all other elements that help us interpret the world and are realized in an encounter with the confronting phenomenon: "To distinguish between a normative function and a cognitive one is to separate what clearly belong together" (ibid., p. 309). Thus, beyond casting her native normative context onto a new technology, the user would simultaneously be confronted with certain moral ideas and inclinations that technological design and the surrounding world suggest. Such a moral dimension of Gadamer's hermeneutics is what makes it particularly useful for this study, which I will elaborate on below.

Gadamer (1975/2004; 1977) was primarily concerned with clarifying how people understand the world[3] and how exactly the interpretation process occurs. His answer was the principle of the *hermeneutic circle*, an adaptation and expansion of the similar idea of Heidegger (1927/1962). People always have ideas about a phenomenon that confronts them to see something as something. Interpretation, then, is a constant movement of recognizing the parts to form a whole picture, for it to only become a part of something new again. The seemingly established ideas are, in fact, dynamic and open to revision once a new situation questions them. Gadamer described this intertwined establishment-and-revision process as circular, whereby the new meaning joins the existing knowledge structures for future interpretations of new experiences. Thus, the circle is never complete, and an established understanding is never stable.

Gadamer's hermeneutics stresses the productive nature of understanding. People are *historically embedded* beings that cannot escape traditions and frames of reference, be it our collective morality, gained knowledge, experience, perceptions, and own normative ideas. Within the historical imbedding, Gadamer distinguishes the essential and productive role of *prejudice*, discarding its modern negative interpretation in favor of its ancient meaning as prior awareness, or pre-judgment (Gadamer, 1975/2004, p. 273). Prejudice thus denotes the cumulative potential of the preconceptions, provisional judgments, and biases that inalienably direct people to the new phenomena. Although it is never possible to completely disregard our prior awareness, it is critical to attempt to recognize and expel existing biases to view a new phenomenon on its own terms. Viewed as such, prejudice enables the dialogue with the confronting phenomena rather than constituting a hindrance to interpretation: "It is not so much our judgments as it is our prejudices that constitute our being" (Gadamer, 1977, p. 9). Combined, the ideas of historical imbedding and prejudice understood as pre-judgment allow an entry into the mindset of another time, place, or object within the hermeneutic circle.

However, the hermeneutic circle account assumes that interpretation is a direct, albeit circular, process: from the interpreter to the world and back (see figure 5.1). For instance, in using DVAs, the world can be represented in many ways: for example, inquiring about the weather, listening to music, contacting other people, or learning about the current events. Yet the technologically facilitated manner in which the world appears for the person to make sense of does not get a place in the hermeneutic circle. DVAs, by shifting the mode of interaction with technologies to speaking, expand human interaction with the world with new opportunities for engagement without the distraction of typing or swiping. They provide the opportunity to manage music playing without exiting the shower, deepen the embodiment of one's home through connecting spaces and devices[4] by voice (e.g., turning on the coffee machine), arrange home security at a distance (e.g., scheduling home lighting), listen to the book while doing the dishes, or connect to other people in the same space while playing games through DVA together. At the same time, the new voice-based interface co-produces the experiences, accompanying perceptions, and responsibilities of accessing the world and others.

A voice-based interface both presents the world in a specific way and requires a certain manner of interaction. A user is not simply listening to the music, but needs to first frame a request to put it on in a specific way (e.g.,

Figure 5.1 Hermeneutic Circle Account of Human-World Interpretation. *Source:* Image created by Olya Kudina.

Hey Google, play a song X by artist Y from the album Z) and say it in a clear way void of accents or sound variation for the DVA to be able to process it. Spatially, the user needs to ensure that the surrounding environment does not have parallel conversations or loud music already on, place the speaker in the room accordingly, count on a stable internet connection, and that the DVA processes the request correctly from the first try. In the hermeneutic circle account, the meaning of the experience of listening to music is uncovered through a dialogue between the user and the artist—the former working out her prior judgments within the expression style of the latter. What is omitted, however, is the technological medium through which this circular process occurs, leaving it a direct sense-making process between a human and the world.

The hermeneutic circle account stresses the productivity for the context that co-shapes a given meaning-making situation. If we consider this point seriously, the material setting forms not only a passive context of accessing and interpreting reality but actively contributes to the process. Without a material dimension of interpretation, the hermeneutic circle does not provide an encompassing account of sense-making. With this in mind, I turn to the field of material hermeneutics and, more specifically, to the postphenomenological account of Ihde, which focuses on that active role of technologies in interpretation.

Postphenomenology and Ihde's Human-World Relations

Since the introduction of information and communication technologies in the 1960s and 1970s, the hermeneutic approach has developed in scope. In particular, the works of Dreyfus (1972; 2001), Borgmann (1984; 1999), Ihde (1979; 1990), and Capurro (2010) have advanced the field of hermeneutics beyond the interpretation of text and have contextualized human interpretation within an increasingly technological environment. Dreyfus (1972) highlighted the importance of information technologies in producing the contexts and practices from which people draw knowledge, framing technologies as annihilating "vulnerability and commitment," essential for interpretation and understanding (2001, p. 102). Borgmann stressed the hermeneutic potential of technologies in enabling or disabling social practices (1984) and fracturing contexts of understanding between the "real" world of people and the "virtual" world of technologies (1999). According to Ihde, reality appears as a technologically mediated product or image to interpret, and by mediating our perceptions, technologies also help to shape our actions in the world. Ihde suggested that this perception-action mediation of technologies calls for an introduction of "thing interpretation" (1998, p. 8), "instrumental visual hermeneutics" (ibid., p. 177), and "a hermeneutics of things" (ibid., p. 187), or in other words, material hermeneutics (Verbeek, 2003). According to Capurro, technology produces invisible networks that people cannot fully control and thus weakens people as the interpreters of the world (2010, p. 36).

While this short account of the recent developments in hermeneutics is by no means exhaustive, it depicts an important shift—the acknowledgment of the significance of technology in the process of making sense of the world and oneself in it. Dreyfus and Borgmann consider technology alienating and a hindrance to understanding. However, they do not specify how to identify a place for the increasing presence of technologies in the interpretative frameworks of people. Capurro, in contrast, withdraws from normative valuations and stresses the ontological nature of technologies, in parallel reducing the productive role of people in the process of interpretation. Ihde's account of technologies as mediators seems to provide a non-reductionist view of the role of technologies in interpretation that may also help to better understand the human appropriation of technologies.

Interpreting the co-shaping of people and technologies philosophically on the axis of perception-action (Ihde, 1993), the mediation approach additionally makes room for exploring how specific moral perceptions and actions emerge in human-technology relations (Verbeek, 2008). Consider how a shift from a written to a spoken interface in the case of DVAs and other virtual agents makes adult users perceive them as more trustworthy and increases their proclivity to share personal information (Nass & Brave, 2005). Similarly, regardless of the outer appearance as a plastic box and the robotic glitches in the voice, Druga and colleagues (2017) showed how the voice-based interface and the unchallenged answering of non-stop questions increase the anthropomorphism tendencies of children toward DVAs, generating trust and authority in the device akin to role models. The hardware and software, type of interface, physical appearance, and algorithmic underpinning all enable specific technological practices, magnifying some aspects of reality or suggesting certain use patterns, while reducing the visibility of the alternatives. The non-neutrality of technological design renders technologies as mediators of our relation to others and the world, while retaining the active role of people and their sociocultural setting in co-producing these mediations (Ihde, 1990), including their moral dimension (Verbeek, 2011).

Ihde (1990; 1993) analyzed the relations between people, technologies, and the world within the framework of postphenomenology. This framework builds on phenomenology in exploring the lives of people but extends it further to account for the ever-increasing intertwinement of human lives with technologies. Contrary to Gadamer, Ihde suggests that human-world relations are not direct because technologies play an active mediating role in how people perceive, interpret, and act in the world. Postphenomenology and the technological mediation approach aim to explore how technologies co-shape the reality of people.

To analyze how technologies mediate human existence, Ihde (1990) distinguishes four types of human-technology-world relations: embodiment, where a technology-in-use disappears from view and becomes a transparent counterpart to the human experiences (e.g., a pair of glasses); hermeneutic relation, where a person interprets the world through a technology (e.g., a computer); alterity, where a person interacts with a technology to access the world (e.g., an ATM machine); and background, where technology forms an active backdrop of human experiences and becomes noticed only when malfunctioning (e.g., Wi-Fi). In what follows, I focus on the hermeneutic relation as the most pertinent to the study of sense-making, bearing in mind that all four relations may be viewed from the material hermeneutics angle[5].

For Ihde, technologies are "perception-mediating and perception-transforming devices" (1998, p. 185). Ihde suggests that perception originates within interconnected yet different "bodies:" "Body One" and "Body Two." *Body One* refers to the sensomotorial physical body, or perceptual bodily awareness, "a being-here, located, sensory being with specific styles of movement" (ibid., p. 89). *Body Two* concerns a lived, social body enriched with the history and culture surrounding it. Examining perception through two bodies mirrors Ihde's earlier conceptualization of perception as consisting of both the micro and macro levels (Ihde, 1990). Ihde's combined view on perception would regard microperceptions as pertaining to the perceptual bodily awareness of Body One and macroperceptions as pertaining to the culturally and experientially informed Body Two. Interpretative activity depends upon the dynamic interrelation of the two bodies, both sensorial micro- and cultural macroperceptions.

The hermeneutic relation represents another caveat to Ihde's hermeneutics, which is typical for all four types of relations: while technologies mediate how people relate to the world, it is unclear how the mediated world finds its way back to people. When technologies mediate what appears to us as real and co-shape our object of interpretation, the mediated interpretation cannot go unnoticed for the reference schemes in our micro- and macroperception that embody all lived experience and prior awareness. Particularly in the context of interpretation, the linear scheme in human-technology-world relation seems problematic.

In his 2005 book *What things do?*, Verbeek expands postphenomenology by asserting that people and technologies co-constitute each other. No pre-given subjects exist who act upon the passive objects in the world: "What the world 'is' and what subjects 'are,' arises from the interplay between humans and reality" (Verbeek, 2008, p. 13). Applied to Ihde's material hermeneutics, the co-constitution principle replaces the relations of linearity with circularity. However, it is not clear how the principle of co-constitution functions. Figure 5.2 below schematically represents Ihde's hermeneutic "human → (technology–world)" relation (1990), which overlaps with Verbeek's co-constitution idea (2005).

Figure 5.2 represents the inclusion of technologies in the sense-making processes of people through the perspective of mediation. The outer world appears to the DVA user not directly but via a gendered voice of the device, commands of the smart speaker, a crafted persona of the DVAs, and its design exigencies (e.g., connection to electricity and Wi-Fi, speaking loud enough, etc.). The world, in Ihde's words, appears as "framed" because "what is presented is presented as already distinct from ordinary or lived-bodily space" by virtue of "limited and selected-out *framing*" and "an on/off presentation" (1998, p. 91, original emphasis). DVAs mediate the perception of the user because they simultaneously reduce the sensory and audio experiences by framing the world through its particular design and modes of presentation and magnifying it by allowing access to the world that is not available to the naked eye or touch.

In figure 5.2, the circular arrows around Ihde's "human → (technology–world)" relation (1990) represent Verbeek's co-constitution idea (2005). Here, the historical and cultural Body Two allows appropriating the socially positioned device and the messages it produces. The sensomotorial Body One, on the other hand, accounts for the physical interaction with the DVA, ensuring the issuing of commands such that the speaker can process them (e.g., language proficiency, tone, speed, clarity) and the optimal functioning of the device for the user (e.g., privacy settings, wake-up word, voice settings, battery charge, etc.), which together contribute to the process of interpretation. As such, both the sensomotorial Body One and the experiential, sociocultural Body Two produce meaning. However, while Verbeek augments Ihde's material hermeneutics by acknowledging the continuous dialogue between the different counterparts of mediation, he does not explain how such a feedback channel functions and leaves the issue of linearity unresolved.

In the case of DVAs, the linearity of human-technology-world relations disembodies the users from their physical bodies (Body One) and their sociocultural setting (Body Two) by precluding the feedback through the DVA among the user and the world. An expansion of voice-based interfaces

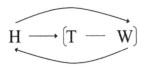

Figure 5.2 Material Hermeneutics Account of Human-Technology-World Interpretation. *Source:* Image created by Olya Kudina.

introduces a new way of experiencing the world, while at the same time constituting specific types of subjects that would fit the technologically mediated spoken interaction and excluding those beyond the category of a standard user. Current speech recognition systems are poorly equipped to process the speech variance and impaired speech intelligibility due to, e.g., oral cancer (Halpern et al., 2020), cleft lip and palate (Schuster et al., 2006), or stroke survivors (Jacks et al., 2019). In the current scheme of human-world relations, a near inability of DVAs to process pathological speech or speech variance breaks down access of certain users to the world, an example of linearity in the absence of further underlying embedding of Body One in the sense-making process. Regarding sociocultural Body Two, consider how the dominant voice assistants on the market assume English language as the average standard, leading to processing errors whenever non-native speakers, people with accents or any language variation interact with DVAs (Wu et al., 2020; Palanica et al., 2019). Pyae and Scifleet identified an inherent bias not just toward English language proficiency but also regarding the culture of the English language that made non-native speakers struggle with DVAs, forcing them to change how they "constructed their mental models and instructed the device" (2018, p. 552). As Sowanski and Janicki suggest, even though the companies gradually expand the language packages of DVAs, the slow pace of the language processing development and the predominant focus on the English language create an unfair fragmentation of society with over ninety-one languages exceeding ten million users with born and acquired speech variations (2020, p. 477). The implicit linearity in human-world relations precludes the larger sociocultural world from updating and balancing out the awareness of Body Two.

These examples demonstrate how technological mediation of the sense-making process influences the subject constitution of the user by not offering a way for the sociocultural Body Two to communicate back with the technology and the world, which affects the user's agency by limiting the feedback from their sensomotorial Body One. Additionally, these examples show how the linearity in the sense-making dimension concerns all four types of human-technology-world relations, not just the hermeneutic one: by disembodying people from the seamless experience of the world, reducing and conditioning access to the world through the DVAs in the alterity relation, and by bringing DVAs from the background of facilitating interaction to an explicit interaction counterpart. Hasse (2008) similarly questioned how material hermeneutics can embed the cultural context in the shaping of specific subjects, ensuring a dynamic, reciprocal exchange between the micro- and macroperceptions on the one hand and technology and the world on the other (p. 47). It seems that when Verbeek (2005) updated the nature and the schematic representation of Ihde's human-technology-world relations, he integrated the unresolved issue of linear sense-making in the hermeneutic, just as in the other types of human-technology-world relations.

If we seriously consider the human-technology-world co-constitution, then, while interpreting the world through technologies, technologies co-shape the prior awareness and understanding of people. Technology enables different or new perceptions that join our bodily and cultural awareness to form a basis for further interpretive processes. For this reason, a person is not the same person and the world is not the same world when they find themselves in a technologically mediated situation. Postphenomenology incorporates technologies in the interpretation process as mediators. However, by not explicating how the mediated world gets embedded in the perceptions of people and how people can then act on them, human-technology-world relations continue to be linear, leaving the human and the world sides of the interpretation process as passive counterparts. Thus, an encompassing account of interpretation is still missing—the one that would acknowledge the active role of people, technologies, and the sociocultural world. The following section attempts to produce such an account, grounded in both Ihde's account of material hermeneutics and Gadamer's account of the hermeneutic circle. It seems that both accounts have something that the other is missing and an opportunity presents itself to combine them, while necessarily updating.

MEDIATED MEANING AS A HERMENEUTIC LEMNISCATE: FROM PEOPLE THROUGH TECHNOLOGIES TO THE WORLD—AND BACK

A broader understanding of the sense-making process than presented in the traditional hermeneutic and material hermeneutic accounts would need to consider the formative awareness of people, the mediating role of technologies, and the productive nature of the sociocultural contexts in the process of interpretation. The structure of such a process of interpretation would resemble a lemniscate (∞), consisting of three linking, interrelated components: human, technology, and the world (see figure 5.3). The technologically mediated *hermeneutic lemniscate* indicates how a sense-making process covers the way people actively appropriate technologies, how the appropriation gets embedded in the world, makes the world meaningful to people in a specific

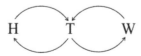

Figure 5.3 **Hermeneutic Lemniscate Account of Human-Technology-World Interpretation.** *Source:* Image created by Olya Kudina.

way, and reconstitutes the subject of technological appropriation in return. Thus, the technologically mediated sense-making, embodied in the hermeneutic lemniscate, indicates how technological appropriation constitutes the world for a person, and against that background, she gets reconstituted as a subject, meaningful in a new way.

In figure 5.3, *H* represents a person enacting her prior awareness to engage with a DVA (e.g., asking about the local news). *T* represents a smart speaker as a technologically mediating actor, suggesting certain perceptions and actions. Finally, *W* represents the sociocultural world, a soundboard against which technology is appropriated and that also gives it a certain meaning. An emergence and temporary stabilization of meaning occur when the process of interpretation passes through all three counterparts. Bearing in mind the relational and intentional aspects of technological use, the hermeneutic lemniscate does not link two separate hermeneutic circles, e.g., (H-T) and (T-W), but rather presents the sense-making process in its interrelated unity: from human through technology to the world and back. Nonetheless, just as in Ihde's postphenomenological model of relations, analyzing the segments of the lemniscate allows elucidating a specific part of the sense-making process and making it available for reflection, minding that all counterparts are interrelated and affect one another.

For instance, to hear the local news, a person must interact with the DVA: turn it on with a special wake-word, choose the news station, and issue a clear command (e.g., Alexa, play news from XYZ) (the upper left curve of the lemniscate). The DVA then applies the user's preferences to its default settings (e.g., GPS location), databases, and options (the lower right curve). Together with these processed preferences, the news piece carries specific messages that DVA will transmit (the upper right curve). The news presentation, however, appears to the user as framed not only by its value-laden content but also by the design features of the DVA: gender of the voice (female by default in many DVAs, with an option to change the settings), the persona behind the assistant (e.g., chatty or formal), portability of the device, quality of the Internet connection, etc. that together shape the news experience and the productive interpretation context (the lower left curve). The going back-and-forth between the initial request of the user, the technological mediation of the speaker and the news, representing an image of the world, constitutes in Gadamer's words a fusion of horizons. This sense-making activity leads to a temporary understanding and a certain appropriation mode of the DVA, subject to revision when the prior awareness changes. An example of this could be parents choosing to remove a smart speaker from their child's bedroom, initially used for a DVA-delegated bedtime fairytale, concerned about DVAs privacy issues discussed on the news or having experienced occasional device glitches themselves. In this case, both the projective (DVAs as a risk) and practical (positioning and use pattern) appropriations of the device change when additional meanings join the interpretative structures.

Crucial for the lemniscatic principle of interpretation, Gadamer's hermeneutics gives it weight to the context, the present situation in the process of interpretation—the sociocultural world is not a mere passive embedding but too can bestow variance to the establishment of meaning. This guarantees that there is no one true meaning as a result of interpretation, but rather that interpretation is always open and relates to a variety of ever-expanding contexts. Additionally, both the interpreter and what confronts her are active parties in the process of interpretation. Such feedback channels between the three counterparts to interpretation showcase the productive blending of several contexts and explain how technologies can always embed multiple stabilities, lending alternative established meanings appropriate to the context (Rosenberger, 2014). In this way, there is no single static meaning to a DVA: the device may be perceived simultaneously as a companion, a narrator, a teacher, a dictionary, etc. that all add to the meaningful experience of being in the world.

The circular principle explains the dynamics of the lemniscate: that productive fore-structures of human understanding set the appropriation of technologies in motion; how this proceeds in circular back-and-forth motion, revising the existing pre-judgments and becoming embedded in the existing structures of interpretation; and how continuous interpretation, always subject to revision in view of the expanding contexts, culminates only in a temporarily stable understanding. This dynamics in the lemniscate, accompanied by the productive engagement of people, technologies, and the world ensures that their relation overcomes the linearity in the account of material hermeneutics.

Note that the fact that a technology finds itself in the middle of the lemniscate does not attribute it with a central, primary place in the process of meaning-making. On the contrary, I suggest approaching it as a fluid process where none of the parties to the interpretation occupies a central role because all three actively contribute to the interpretation. This also means that a human is not the default point of entry into the lemniscate, as suggested by Gadamer's hermeneutics. One could conduct an appropriation study beginning with a technology (e.g., to discover its embedded scripts and affordances) or with the specific sociocultural context (e.g., to understand why household adoption of DVAs might go faster in Asia than in Europe). In this chapter, I started exploring the appropriation of DVAs from the human side because I wanted to understand how people make sense of technologies while making sense of their environment and themselves. However, the lemniscate principle of interpretation would always simultaneously inquire what the technology and the larger cultural world project and how this reflects on the person in question. What one examines is a specific hermeneutic situation that must be studied from all sides, regardless of the initial entry point. As such, human, technology, or the world cannot be singled out as the center of the lemniscate.

The lemniscate refines Gadamer's hermeneutic circle account by foregrounding the active role of technologies in the process of interpretation. On the other hand, the continuity and fluidity of the lemniscate also clarify Verbeek's co-shaping idea, explaining why the interpretative structures of human understanding are never static and how the technologically mediated world returns to people. Namely, the effective histories of people ensure that the initial meaning of a technology remains in flux. As soon as a preliminary meaning is established, it becomes the background for further interpretation, triggered by the new information about the device, the expanded context of the application and the user experience. The process of appropriation is never finished. This also ensures that the way people use technologies can never be limited to suggestive design, giving people the freedom to enable multiple stabilities and interpretations of the same technology.

MORAL HERMENEUTICS AND TECHNOLOGIES

The hermeneutic lemniscate also allows zooming in on the moral mediation of technologies by showing how specific value manifestations, tensions, or moral opportunities arise in the encounter of people, technology, and the world. The hermeneutic lemniscate embeds this technologically mediated moral encounter between the user, technology, and the sociocultural embedding. Following the technological mediation approach, this encounter can proceed along the lines of co-shaping moral perceptions, actions (Verbeek, 2011), and values of people (Kudina, 2019).

For instance, consider how the current generation of DVAs co-shape *moral perceptions* on the role of women. DVAs predominantly have female voices and by design may not retaliate to rude behaviors, facilitating an image of women as docile obedient servants who can never refuse a request. As Fessler (2017) found through her insightful ethnographic study, DVAs reply to explicitly sexualized offenses and requests in a playful indirect manner, for instance, with Apple's Siri responding to a statement "You are hot" as "How can you tell? You say that to all the virtual assistants." Even though the intention behind choosing female voices for the DVAs might have been to make users more comfortable to talk with a machine (Nass, Moon, and Green, 1997), it in parallel contributed to reinforcing gender stereotypes. Specifically, in their UNESCO report, West, Kraut, and Chew (2019) suggest that a pervasive spread of predominantly female-voiced digital assistants "reflects, reinforces and spreads gender bias," propagates "tolerance of sexual harassment and verbal abuse," creates a perception of women as "the face and voice of servility and dumb mistakes" among other issues (pp. 86, 104-115). DVAs illustrate how moral intuitions get shaped not only through the content of interaction but also by the technologically mediated manner in which it is presented.

Moral actions can also get direction through the use of DVAs, for instance, in the case of privacy. The privacy concerns regarding the DVAs often refer to the third-party listening in on what users say to their voice assistants and the fact that the DVAs will not work unless their microphones are in the always listening mode (e.g., Bugeja, Jacobsson & Davidsson, 2016; Lau, Zimmerman & Schaub, 2018). In contrast, I would like to draw attention to how DVAs mediate the privacy attitudes by shaping novel social practices. Regarding DVAs, the privacy concerns are often framed as an inevitable trade-off: the user either accepts the privacy risks of placing a smart speaker in their home to be able to continue to use the device, or refuses to use it altogether. Such a dichotomy need not be warranted.

Recognizing how technologies participate in our moral sense-making and co-shaping our moral inclinations and actions allows us to develop alternative ways of using them that both support our values and do not exclude using the technology. Some creative appropriations of smart speakers include building a muffin-top cover that continually distracts the speaker by feeding white noise into it, designed by the Project Alias (2018). The speaker will only hear the users when they use a special wake word that they themselves chose for the speaker. Some people developed contextual uses for smart speakers, unplugging them at certain times during the day. Yet others choose not to use them anymore. What these efforts show is a recognition that DVAs do something more than what they were designed to do, and they are trying to explore what that "more" means. Such creative user efforts represent an active way to shape what is meaningful regarding privacy when using DVAs without the either-or trade-offs.

Finally, the *values* themselves can *undergo formation and revision* in the process of using smart speakers. Specifically, the value of good and meaningful communication becomes at stake in the DVA-mediated interaction. DVAs frequently mishear and wrongly process the speech of adult users, especially those who are native to languages other than English and have accents (Wu et al., 2020). As a result, such users treat DVA errors as limitations in their own language, leading them to plan and adapt their speech in interactions with DVAs (ibid., p. 9). Burton and Gaskin (2019) showed that adults in general generate frustration and anxiety in interaction with DVAs, even though barking orders at the device does not make them impolite in human-to-human interaction. The authors suggest that because children's behavior patterns are not as formed as in adults, interacting with DVAs can incentivize them to negative socializing guidelines.

An empirical study of children's adoption of DVAs by Ureta and colleagues (2020) demonstrated how "the short waiting time, frequent interruptions during pauses and mishearing words" (p. 501) negatively influences children's speech and their cognitive processes. Wiederhold (2018) concluded that because DVAs do not have any reprimanding features as human interlocutors and encourage short command-based interactions, children learn to expect instant gratification, e.g., immediate answers to questions and entertainment. Building on some of these concerns, Bonfert and colleagues (2018) designed DVAs as explicit role models that would actively rebuke discourtesy and reinforce polite behavior by encouraging words such as "Please" and "Thank you," a move that Amazon also took in 2018 when it piloted its Kids Edition of the Echo smart speaker. While the empirical studies on the adoption of DVAs continue to surface and their global introduction is still ongoing, the hermeneutic lemniscate can point to how the value of meaningful communication gets shape in the interaction with DVAs and its specific mediating features, suggesting areas of moral attention for its responsible design and use. For instance, regarding children, is it desirable to delegate the enforcement of polite interaction to DVAs, even if it is technically possible? On a different matter, as Wiederhold suggests, "Instead [of instant gratification], it is important for children to experience the discomfort of confronting an obstacle, the triumph of problem-solving, and the ability to recognize their own autonomy" (2018, p. 471). Designers can consider storytelling and dialogue techniques to engage children in multi-agent interaction patterns, sensitizing them to the accountability and sociality that goes with it (Ureta et al., 2020). Thus, DVAs promote further parental responsibility and require intervention early on, while at the same time making it tempting to delegate some of the parental duties, such as patiently answering non-stop detailed questions, playing word games, or telling a bedtime story.

Considered within the framework of hermeneutic lemniscate, these examples illustrate how the moral landscape of people is dynamic and responsive to the technological practices. The lemniscate model can explain the interpretative processes that enable value dynamics by unraveling the interrelation between the productive background of human interpretation, the technology with its particular mediations, and the active sociomaterial context that shapes the given practice. The hermeneutic lemniscate thus sheds light on the intimate relationship between values and technologies, showing how the normative context can be reconfigured throughout the human appropriation of technologies.

CONCLUSION

In this chapter, I have argued that technologies, such as AI-powered voice assistants, actively participate in the moral sense-making. I have relied on the hermeneutics of Gadamer and the material hermeneutics of Ihde and Verbeek to develop a lemniscatic principle of technologically mediated sense-making. The lemniscate principle shows how people, technologies, and the sociocultural world actively participate in the formation of meaning by engaging the productive fore-structures of human understanding, setting the appropriation of technologies in motion, and constituting a meaningful world for people, where they in turn get reconstituted as specific subjects. The process of technological appropriation in parallel activates the moral sense-making that forms an inevitable part of our interpretative structures and that we enact when making sense of technologies. Using DVAs as an example, I have shown how these AI-guided devices can mediate our moral inclinations, decisions, and even values.

Understanding the workings of the lemniscate and the specific mediating roles of technologies makes informed use and design a cornerstone to living well with technologies. Technologies introduce new practices, situations, and choices that make us aware of what is important to us. Even though such new practices may challenge our existing values, as the examples I have analyzed in the chapter suggest, people continuously prove their inventiveness to actively shape what is important to them. Beyond this active user responsibility, design practices can facilitate an active engagement with technological mediations.

Designers can—and already do—anticipate the future and untangle current appropriation modes to make the final product relatable to the users in a variety of contexts. Acknowledging the mediating nature of technologies in the sense-making additionally helps to avoid the artificial dichotomies, as discussed in the privacy either-or model of DVAs. Similar to devising a cheap and efficient privacy add-on to the popular voice assistants (Project Alias, 2018), designers and researchers developed Q, a genderless voice model for DVAs, challenging yet another duality (Genderlessvoice.com, 2019).

Treating technologies as mediators of our sense-making also implies a new type of user responsibility: to explore the practice a technology creates, to inquire what its design and interface promote, which actions become less possible, how it shapes the interaction with others, whether any value conflicts or opportunities appear, and why. The hermeneutic lemniscate allows multiple entry points to examine the daily practices with technologies. It also helps to keep an open answer to the question "Alexa, who am I?," one of the first questions the Amazon DVA users ask the device to help it recognize their voice, but that inevitably has a deeper meaning, especially in the context of technologically mediated interpretation. Some of the questions for further research would be inquiring whether the hermeneutic lemniscate allows examining the sense-making process not just at the individual level, but extending it in the wider social and political realm, and if so, how. Additionally, specifically in relation to AI-based technologies, it would be interesting to address how the lemniscate model simultaneously produces multiple human-world relations enabled by the self-teaching algorithms, learning about the user behavior and the social tendencies and producing new forms of use.

In this chapter, I have synthesized the findings of the earlier chapters to offer the hermeneutic lemniscate as one way to explore the non-reductionist interrelation of people, technologies, and the sociocultural setting through a joint production of meaning. Beyond additional responsibilities for design and use, analyzing in this way how everyday technologies mediate the way we understand ourselves, each other, and the world in parallel enables us to use them in an informed way. The following and last chapter will contextualize the preliminary conclusions offered here within the fields of responsible design, introduction, and governance of technologies. It will provide concrete suggestions on how to consciously shape what is important to us, individually and collectively, while accounting for the moral hermeneutics that underlies our decision-making whenever technologies are concerned.

NOTES

1. The sections of the chapter that follow appeared in the open-access article by the author: Kudina, O. (2021b). "Alexa, who am I?": voice assistants and hermeneutic lemniscate as the technologically mediated sense-making. *Human Studies, 44*(2), pp. 233-253.

2. See, for instance, the works by Reijers, Romele and Coeckelbergh (2021), as well as by Reijers and Coeckelbergh (2020) on Ricoeur's hermeneutic scholarship for philosophy of technology.

3. Although Gadamer primarily referred to textual interpretation in understanding the world, his hermeneutic account has been applied in a broader sense to any particular object in the world (Fry, 2009).

4. There is a range of third-party household appliances that can be connected through internet to DVAs, e.g. a bathtub, a thermostat, a coffee-machine, security systems, etc.

5. Relating Ihde's broad definition of material hermeneutics to the fourfold scheme of human-technology-world relations means that each of these relations can be viewed from the angle of material hermeneutics. The hermeneutic relation would explicitly review the technological mediation of interpretation. The embodiment relation would inquire into how the person reveals herself to the world and the world to her through incorporated technologies. The alterity relation with hermeneutics in mind would be enabled when the technological design communicates certain practices with this technology. Finally, the background relation would uncover the silenced blending of technologies with our environment.

Conclusion

Reflecting on the Moral Hermeneutics Study from the Perspectives of Technology Design and Governance

In this final chapter, I will reflect on the overall ambition for this book, namely understanding how technologies mediate human values in the study of moral hermeneutics, and reflect on its implications for the fields of responsible design and governance of technologies. The interrelation of values and technologies was my point of departure. In what follows, I will conclude the book by referring my findings on technologies and moral sense-making back to the discussion on the ethics of technology, as well as sketching the avenues for further exploration. To this end, I first review the potential contribution of the technological mediation approach to the practical field of technology design, more specifically, its relation to the method of Value Sensitive Design (Friedman, 1999). More specifically, I explore how the technological mediation account expanded with the line of moral hermeneutics can contribute to it. Afterward, I turn to the field of responsible technology introduction and governance to understand what the moral hermeneutics study can imply for it. To do so, I reintroduce the ethical variant of the Collingridge dilemma that I briefly discussed in the Introduction. More specifically, I analyze the dilemma in view of the book's findings and correlate it with the existing ethical approaches that deal with technologies and values, earlier discussed in chapter 2, namely the sociotechnical experimentation approach of Van de Poel and the technomoral change approach of Swierstra. In combination, this will allow me to draw conclusions regarding the contribution of the moral hermeneutics account, reflect on its limitations, and sketch directions for further research.

DESIGNING TECHNOLOGIES WITH THE MORAL HERMENEUTICS IN MIND

Various design approaches attempt to address in practice value-laden technology design, incorporating ethical reflections into the development of technology early on (e.g., Value Sensitive Design, or VSD for short [Friedman, 1999], Care-centered VSD [Van Wynsberghe, 2012], Disclosive Computer Ethics [Brey, 2000], Designing for Trust [Camp, 2003], Design for Values [Van de Poel, 2015; Van den Hoven, 2005], Values at Play [Flanagan, Howe & Nissenbaum, 2005], Product Impact Tool [Dorrestijn, 2012]). However, VSD, piloted by Batya Friedman and colleagues in the 1970s, stands out as one of the earliest, most comprehensive and, to date, most influential approaches attempting to methodically incorporate ethical considerations regarding technologies in the design process. For these reasons, in the remainder of the chapter, I focus on this approach to review its take on values and the considerations of moral hermeneutics in design.

Value Sensitive Design belongs to the practical domain of ethics and posits that the normative ideas of stakeholders inform the process of technology development (Friedman, Kahn & Borning, 2002; Friedman & Kahn, 2003). As such, technological design is not neutral, and VSD thus aims to account systematically for the human values present within it from the early stages. To accomplish this task, VSD researchers utilize iterative, three-dimensional investigations: conceptual, empirical, and technological. In the conceptual stage, researchers map out the stakeholders, the values in play, and whether value conflicts occur. This is followed by an empirical investigation into the design and use context to determine how the identified values are realized, which of the conflicting values receives priority, and why. At the level of technical investigation, a design system is built to implement and support the findings of the conceptual and empirical stages and accommodate all pursued values. In this way, VSD enables a win-win situation by means of a theoretically grounded design approach to accommodate different conflicting values for the benefit of all stakeholders, and the method has evolved to incorporate a wide variety of practical sub-methods of VSD applications with the value considerations in mind (Friedman & Hendry, 2019).

However, to be able to design the identified values into a technology, these values must be specified in concrete design requirements. To achieve this, Friedman and Kahn (2003) suggest conceptualizing values in a generally accepted manner to ensure that VSD can accommodate different value manifestations from varying contexts into a universal value set analysis for design. As such, the authors endorse a set of twelve "human values with ethical import[1]" (Friedman & Kahn, 2003, p. 1187) that incorporate both the traditional moral values of rule and consequentialist ethics (such as autonomy, freedom, or trust) and more modern values inherent to the Information and communication technology (ICT) community (such as ownership, calmness and environmental, sustainability). This commitment was further extended when Friedman and colleagues (2006) embodied the identified value set in a guideline table on value meaning and specification to be considered in design. Although the authors maintain that this list of values is tentative, flexible, and subject to ad hoc scrutiny, there nonetheless exist several problems worth considering in detail.

The primary issue, which yields the subsequent ones, addresses the approach to values in VSD. At first glance, the VSD founders plead for a middle ground between a set of universal and particularistic values, affirming VSD as an approach "that allows for an analysis of universal moral values, as well as allowing for these values to play out differently in a particular culture at a particular point in time" (Friedman & Kahn, 2003, p. 1183). At a closer look, however, a clear priority in design is attributed to the preselected set of twelve values mentioned above, with a later addition of a thirteenth, the value of courtesy (Friedman et al., 2006). Moreover, the VSD founders "highlight [the] ethical status [of these values] and thereby suggest that they have a distinctive claim on resources in the design process" (Friedman & Kahn, 2003, p. 1187). Repeatedly emphasizing in a later work that the suggested list is not exhaustive, however, the authors elaborate that the list is "intended as a heuristic for suggesting values that should be considered in the investigation" (Friedman, Kahn & Borning, 2006, p. 366). As a result, while VSD aspires to provide a method applicable to values in design in general and promises to reflect a middle ground between universal and particular takes on values, *de facto*, VSD approaches the extreme of universalism. This makes it vulnerable to consequent issues of detachment from the practices and lived experiences of people, generalizations, and unspecified origins that I have elaborated upon in chapter 2.

Although VSD has received recognition and merit from many academics and design practitioners due to its desire to include ethics in the design process and a comprehensive method to do so (e.g., Cummings, 2006; Timmermans, Zhao & Van den Hoven, 2011; Van de Poel, 2009b; Van den Hoven & Manders-Huits, 2009), it has also received a share of criticism, often related to the suggested value list (e.g., Alsheikh, Rode & Lindley, 2011; Borning & Muller, 2012; Le Dantec, Poole & Wyche, 2009; Manders-Huits, 2011; Yetim, 2011).

Le Dantec et al. (2009) challenged the choice of a value list as a bold heuristic for being itself embedded in cultural and professional views. The authors demonstrate that the VSD "manifesto of values" arises from the ICT community, which has traditionally promoted "personal expression and collaboration" as well as other cultural commitments that the authors trace to the revolutionary U.S. ideals of the 1960s (ibid.). By devising a list that can claim primacy in design consideration, the authors claim that VSD positions itself "within the nimbus of morality, cultivating a dogmatic response with respect to which values are worthy of consideration and disengaging from a commitment to understanding the nuanced manifestation of a plurality of values" (p. 1142). Moreover, the methodological hierarchy of VSD, argue Le Dantec et al. (2009), supports the view that certain values deserve primary consideration in design. In principle, the tripartite methodology promotes an iterative and mutually informing study on values in design, by virtue of which "an artifact . . . emerges through iterations upon a process that is more than the sum of its parts" (Friedman, Kahn & Borning, 2002, p. 2). However, as the authors note, the ordering of conceptual investigations over the empirical ones, backed by a pre-given value list, discourages practitioners from contextualized and extensive value discovery and value specification.

Mirroring and elaborating on this issue, Borning and Muller (2012) caution that providing a list of values for primary consideration and supplementing it with pre-given conceptualizations can evoke bias in conceptual investigations. They argue in favor of value discovery among stakeholders, which would return the plurality of values to the fore. Another plea concerns the deeper contextualization of the suggested value set, or as the authors assert, "Be explicit about the particular culture and viewpoint in which they [the suggested values] were developed, rather than [. . .] making implicit claims about more universality than are warranted, or perhaps even intended" (ibid., p. 1126). The crux of the criticism concerns the non-neutrality of the selected value set and the lack of its context qualification. Thus, the incoherence regarding proclaiming and maintaining a middle ground in values entails multiple practical problems that challenge the integrity of the VSD approach.

This leads to the deduction of an overarching normative challenge to the VSD approach, which can be formulated as "Whose and which values in design?" Alsheikh and colleagues (2011) address the question of value bias in VSD with a case study on long-distance relationships mediated by ICT in an Arabic cultural context and invite VSD to be cross-culturally sensitive to "understand users in terms of their own values and priorities" (ibid., p. 83). Manders-Huits (2011) instantiates this issue and questions VSD's take on identifying values and translating them into a final product. She also argues for a more contextualized value conceptualization, cautioning, however, avoiding the trap of the naturalistic fallacy of conflating empirical facts with ethical values, or "by reducing an 'is' to an 'ought'" (p. 279). Echoing the voiced concerns and addressing the potential issue of a naturalistic fallacy, Van Wynsberghe and Robbins (2014) argue for "a proper value analysis," providing an ethical reflection on the identified list of values, embedding them in context, and "correctly translat[ing] values into contemporary norms" (albeit ultimately saying that this is a tall order for engineers and designers and proposing trained ethicists as more suitable candidates for the job) (p. 956). Overall, the critics of VSD ask for reflective value discovery and conceptualization, sensitive to the value context.

The perspective of moral hermeneutics in the technological mediation approach could provide guidance on how to address the broad issue of value discovery and conceptualization in VSD. The primary difference between the two approaches concerns the position on the nature of values considered in design. The VSD approach endorses the view that technological design allows for specific "value suitabilities," where "a given technology is more suitable for certain activities and more readily supports certain values while rendering other activities and values more difficult to realize" (Friedman, Kahn & Borning, 2002, p. 61). This view partly correlates with the basic tenets of the technological mediation approach, whereby technologies are intentional, in that they transform how people perceive and act in the world, readily presenting one set of options to the user, while foreshadowing an alternative one (Ihde, 1990). However, the concept of "value suitabilities" in VSD remains unproblematized because of the passive correlation VSD attributes to values and technology. While in VSD, values are considered as an *ex-ante* or *ex-post-facto* addition to the design process, from the mediation perspective, values, like people, do not exist outside, out there, but are the outcome of mutual shaping with technology in the practices and experiences of people.

Following the technological mediation approach, not only can technology amplify and reduce certain facets of the world, but by doing so, it also enables a distinct reality, which people perceive and upon which they make practical decisions; or, as Verbeek (2006) contends, "When technologies co-shape human actions, they give material answers to the ethical question of how to act" (p. 361). For this reason, it is also important in the process of technological design to understand how values are constituted and negotiated in relation to technologies. In view of this, a mediation analysis of moral sense-making, as I have shown in chapters 3 and 4 on the human appropriation of technologies, places much attention on studying the practice and context within which a technology is or will be embedded and understanding how what is meaningful and valuable to people takes shape in that process. With these considerations in mind, I suggest that the moral hermeneutics study, building on the technological mediation approach, could enhance the VSD approach, the stage of both the conceptual and empirical investigations, with the missing value qualifying and contextualizing component.

Sensitivity to values remains a core benefit of the VSD approach. It manifests in considering the design process as non-neutral and imbued with values. However, this sensitivity is largely lost in the process of translating values into norms and design requirements, since concrete design options require concrete value specifications.

For instance, the value of privacy frequently draws attention from VSD scholars. Most often, it is conceptualized under the lines of control of information, with informed consent as a primary mechanism to support such a value of privacy (e.g., Friedman et al., 2002; Van den Hoven & Vermaas, 2007). While following narrow specification guidelines makes ethics in design operational, such an approach also risks embodying only certain aspects of the value, ignoring the complex context within which the value is predominantly manifest. The findings of the exploratory study on Google Glass and privacy in chapter 3 empirically support the theoretical suggestions, urging a turn to the practices of people and the anticipated application context.

Van Wynsberghe (2012), in her study on healthcare robots, identified a similar problem regarding a narrow take on values in VSD. According to her, the VSD approach must more closely consider "the context, the practice, the actors involved and how care values are manifest" (p. 120). Van Wynsberghe consequently developed a care-centered VSD approach that evaluates the practice regarding the potential impact that the introduction of care robots could have on it. An appropriation study grounded in the mediation approach could further enhance VSD to make it applicable for a variety of contexts and technologies.

The integration of the mediation lens with the stages of conceptual and empirical investigations in VSD can enable the developers, designers, and engineers of technology to better comprehend values in design and to enhance reflexivity during conceptual and empirical investigation. The moral hermeneutics analysis could primarily be helpful to deepen an understanding of the value in design by studying how it manifests in current or anticipated use practices, as even an exploratory mediation study of privacy and Google Glass revealed multiple value dimensions and design considerations that could potentially be used for a redesign of Glass. Equally important, and in line with the value-laden ideas of technological design that Friedman and colleagues endorse, is to understand how VSD practitioners themselves attribute meaning to this value in relation to a technology under construction. Here, an Interpretative Phenomenological Analysis study explores how people-appropriate technologies can help to identify, contextualize, and analyze the moral background of the design problem, which could help to better inform the design process. Together, this might allow VSD to better relate to potential users and application contexts.

Speaking of contexts, as I have demonstrated in this book, technologies also mediate the moral sense-making in the broad dimension of culture, spanning the diverse human traditions and norms, institutional and legal frameworks. In relation to the sociocultural context, technologies reveal different value dimensions. The VSD method aspires to be designed with the different cultural frameworks in mind and wants to be responsive to the changes in the worldview of technology users (Calo et al., 2020). But how can one enhance the VSD method with the moral hermeneutic considerations when culture is concerned? The appropriation study, guided by the lemniscate principle and manifested in the IPA method, can help reveal how technologies mediate values in both existing and anticipatory uses. Mediation findings can accompany designers in their exploratory stages. The technological mediation approach, as expanded with the study of moral hermeneutics in this book, can open the VSD approach to the idea of value change that comes from the side of different world embeddings, as well as offer a practical manner to expand the stages of its conceptual and empirical investigations. It can thus complement it and somewhat account for its existing challenges.

This is not to say that complementing the VSD approach with the moral hermeneutics study would come without its own challenges. One of the questions is who would be responsible for conducting a technological appropriation study. As the VSD method suggests, ethical reflection is not external to designers; it forms an integral part of their work. Hence, I do not suggest incorporating an embedded ethicist on a design team but instead trust that designers know best how to complete their own work. As such, I opt to support the ethical ambitions of designers by providing them with technological mediation tools to streamline the reflection on values in design.

In practice, the scope and depth of the appropriation study and the mediation analysis as I have presented them in this dissertation would require additional training to learn the research process and methodology. Even though the mediation analysis can provide a better-informed background for conceptual and empirical analysis in VSD, designing for a multidimensional value could still require trade-offs, albeit better-informed ones. Moreover, some ethical questions offer no structural way to address them. Designers would have to decide pragmatically, on an ad hoc basis, whether the richness of information the mediation analysis brings would justify the additional time and other resources it requires. Overall, the process would be challenging but feasible. This challenge is not dissimilar to the one that surfaces when considering the moral hermeneutics in the governance of technologies, to which I turn in the following sub-section.

MORAL HERMENEUTICS AND TECHNOLOGY GOVERNANCE[2]

The field of responsible introduction and governance of technology frequently addresses anticipation and foresight regarding the potential use and ethical implications of new technologies. However, because technological implications manifest themselves when the technologies in question are already deeply embedded in practice (Collingridge, 1980), informed anticipation regarding new technologies is a complex affair. The sub-section reintroduces the Collingridge dilemma initially presented in the introduction and relates it to the phenomenon of moral hermeneutics. More specifically, I explore the fitness of a mediation-based moral hermeneutics study to address an ethical variant of the dilemma. I do so in relation to notable approaches in the ethics of technology that deal with value dynamism and change, primarily Van de Poel's sociotechnical experimentation approach (2013) and Swierstra's technomoral change approach (Swierstra, Stemerding & Boenink, 2009).

In the introduction, I briefly introduced a classical dilemma in technology governance, the so-called "Collingridge dilemma" (1980). When a technology is in an early stage of development, it is still possible to influence the direction of its development, albeit without knowing how it will affect society. Yet, when the technology has become societally embedded, we can ascertain its implications, but it remains difficult to influence its development. This dilemma is one of the most significant challenges for the responsible design, use, and governance of technologies.

Various strategies have been developed to escape it. Some strategies focus on anticipation, or "prospective evaluation" (Grunwald, 2009, pp. 1124–1125), to get in touch with the potential future impacts of a technology at a moment when they can still be addressed through processes of technology development. A good example is the approach of Constructive Technology Assessment, which conceptualizes technological development in evolutionary terms and approaches innovations as "variations" that are exposed to a "selection environment" of markets, laws and regulations (Rip, Misa & Schot, 1995). This technique aims to create a "nexus" between variation and selection by anticipating the future implications of technologies during their development.

An opposing type of strategy focuses on regulating the process of innovation rather than anticipating its outcomes. The approach of sociotechnical experimentation (Van de Poel, 2013) is a useful example here. Rather than anticipatively looking into an uncertain future, Van de Poel proposes accepting this uncertainty and approaching innovations as "social experiments" that require ethics to be conducted responsibly. Technologies inevitably change society, and rather than taming this uncertainty by attempting to predict the future, we should responsibly regulate innovation processes.

I wish to explore the role of the technological mediation approach, supported by my findings regarding the lemniscate principle of interpretation and an empirical philosophical appropriation study, as a complementary strategy to deal with the Collingridge dilemma. The findings of this book regarding moral hermeneutics offer another twist on the dilemma, stemming from the idea that technologies co-shape the value frameworks we use to evaluate them. This situation results in an ethical variant of the Collingridge dilemma: when technologies influence value frameworks, the ethics of technology always seem to be either "too early"—evaluating technologies without knowing how the frameworks of evaluation themselves might develop—or "too late"—grasping the ethical impact of a technology but doing so at a moment when the technology has become less prone to change. Or, phrased differently, when we develop technologies on the basis of specific value frameworks, we do not yet understand their social implications, but once we do know these implications, the technologies might have already altered the value frameworks for evaluating these implications.

This connection between technological innovation and moral hermeneutics has a central place in two contemporary approaches in the ethics of technology: Tsjalling Swierstra's approach of "technomoral change" (see chapter 2) and Ibo van de Poel's approach of "sociotechnical experimentation," which was mentioned above. The technomoral change approach identifies the soft qualitative impacts of technologies and develops scenarios to anticipate how technologies influence moral frameworks to inspire deliberation regarding technological practices and policy-making (Swierstra, Stemerding & Boenink, 2009). The sociotechnical experimentation approach takes a radically different direction (Van de Poel, 2013). It considers anticipation too speculative to be reliable and instead approaches technological innovations as "social experiments" that must be conducted responsibly.

However valuable and important these approaches are, they cannot fully address the ethical variant of the Collingridge dilemma. While the relation between technological innovation and moral developments is the explicit focus of technomoral scenarios, it plays only a background role in sociotechnical experiments. Yet, responsible sociotechnical experiments cannot function without an idea of potential future ethical frameworks regarding technologies, and therefore it seems to throw the child with the bathwater to give up on anticipation at large. At the same time, technomoral scenarios can only offer "controlled speculations" about the future, with their ultimate goal being to provoke present-day reflections on a given technology. While sociotechnical experiments embody a piecemeal approach that allows for regulation without speculation, technomoral change relies on scenarios to calibrate technological governance and provoke reflection through enhanced imagination.

I suggest that the technological mediation approach, guided by the lemniscate model through an appropriation study, can complement the two above approaches in dealing with technology-induced value dynamism and the consequent ethical variant of the Collingridge dilemma. As the appropriation studies in chapters 3 and 4 empirically demonstrate, the lemniscate model can make visible and available for reflection the way in which technologies already co-shape moral frameworks. Studying through IPA or any other empirical method how people—often implicitly—articulate new value meanings, ethical conflicts, and dilemmas when discussing technologies, makes it possible to develop a modest and empirically informed type of anticipation regarding how the moral hermeneutics directions may unravel, as an alternative to both technomoral scenarios that outline value change in distant futures and the lack of anticipation in sociotechnical experiments.

Technologies influence human values. The introduction of the birth control pill has changed value frameworks regarding sexuality because it has loosened the connection between sex and reproduction, allowing for new valuations of homosexuality (e.g., Mol, 1997, p. 8). Furthermore, the introduction of Google Glass, as I have demonstrated in chapter 3, reveals multiple dimensions of privacy and precludes an insistence on one dominant meaning. How can we deal with technologically-induced moral hermeneutics in a responsible manner? To answer this question, I discuss and analyze the approaches of technomoral change and sociotechnical experimentation and contrast them with the approach of technological mediation.

Technomoral Change

The central claim of the technomoral change approach is that normative frameworks are not static but instead co-evolve with technologies (Swierstra, Stemerding & Boenink, 2009). The phenomenon of technomoral change should be considered an element of the "soft impacts" of technologies: subtle, technology-inflicted shifts in society, such as changes in user practices, responsibilities, and value frameworks. Often, technology assessment methods and policy-making focus on "hard impacts," such as health risks, environmental security, and economic losses, that can be quantified and often call for yes-or-no answers. In contrast, soft impacts "do not fit well within a techno-scientific discourse [because] they are easily dismissed as romantic, irrational, subjective or vague" (Haen, 2015, p. 21). Yet, the fact that they are difficult to trace does not reduce their importance. For instance, consider the soft impacts of the cell phone, which has enabled people to make phone calls everywhere, experiencing whom we are calling as "closer" than the people physically nearby. This has changed the social acceptability of having private telephone conversations in public. Also, the normative expectation has arisen that people are available to connect anytime and anywhere.

"Technomoral scenarios" can be used to analyze and anticipate soft impacts (Swierstra, Stemerding & Boenink, 2009; Boenink & Swierstra, 2015). A technomoral scenario is a structured method to anticipate soft impacts based on empirical research and analyses of the current practices that might be affected by new technologies. "Emerging technologies, and the accompanying promises and concerns, can rob moral routines of their self-evident invisibility and turn them into topics for discussion, deliberation, modification, reassertion" (Swierstra & Rip, 2007, p. 6). Such new, problematic situations create frictions and destabilizations: conflicts emerge, and values and norms are contested and compete with each other, because they can no longer respond adequately to new problems. It is precisely these alternative destabilizations, with their consequent soft impacts, that technomoral scenarios attempt to foreground to trigger critical reflection regarding the introduction of new technologies.

Because technomoral scenarios function as a deliberation tool to enhance ethical reflection about technologies, they must appear not as far-fetched technological predictions but as controlled speculation about the co-evolution of technologies and morality. At the same time, technomoral scenarios are often set in distant futures to better demonstrate the manifestation of value change. To this end, much empirical work goes into constructing them. For instance, Boenink, Swierstra, and Stemerding (2010) introduced a three-step framework for ensuring the plausibility and empirical grounding of scenarios. To construct the scenarios, researchers must (1) position the emerging technology in the current moral landscape; (2) determine which moral destabilizations the new technology can induce; and, based on (1) and (2), (3) produce a list of potential soft impacts as plausible moral closures that will form the basis of a scenario. To further these efforts, Lucivero (2012) assessed the plausibility of technological promises and concerns to make scenarios into "grounded explorations" of technomoral futures. Haen (2015), relying on the elements of the Conversation Analysis and Discursive Psychology method (te Molder, 2008), studied the politics of public conversations about technologies that traditionally disregard soft impacts as illegitimate. He used technomoral scenarios as a reflection tool in stakeholder deliberations to demonstrate to the participants the morality of their conversations (i.e., how they claim epistemic rights, and accept or reject responsibilities and the effect this has on shaping the technological discourse). Although this is a non-exhaustive representation of the empirical efforts in the technomoral change approach, it demonstrates the evolving concerns and empirical dedication of this method.

The technomoral scenario method helps anticipate the potential social and cultural implications of emerging technologies by provoking ethical reflection in the present. It does not yet, however, offer a method to study technomoral change "in the making,"[3] because it does not address the *dynamics* of the interaction between technology and morality itself but rather its potential *outcomes*. Its empirical pursuits, which are largely directed at making the scenarios plausible and believable in view of the anticipated value change, have mirrored the focus on deliberation regarding the potential soft impacts of technologies. Although the technomoral scenarios approach allows for the enhancement of the quality and content of technological deliberations by drawing attention to the soft impacts, it does not yet allow for the understanding of the dynamics that underlie the value change. To accomplish this, as I further elaborate, the technological mediation approach can provide an empirical basis for studying moral hermeneutics in the present, to complement the technomoral scenario approach.

Sociotechnical Experiments

An alternative manner to deal with the Collingridge dilemma in the ethics and governance of technology was suggested by Ibo van de Poel, in his approach to technological innovation as "social experiments" (2011; 2013; 2016). The central observation behind this approach is that we can never adequately predict the societal impact of technological innovations. The anticipation of societal impacts can only be an adequate way to deal with the Collingridge dilemma when it offers a strong basis for decision-making (Van de Poel, 2016). The wide range of unexpected social impacts of smartphones and the unforeseen risks of the Fukushima nuclear power plant (e.g., Van de Poel, 2011, p. 287) illustrate this. While technologies have the potential to "seriously impact society, for the good as well as for the bad" (Van de Poel, 2016, p. 667), we can hardly predict what these impacts will be. For this reason, according to Van de Poel, we must deal with innovations as "social experiments," interventions in society with unknown outcomes. However, the unknown character of these outcomes does not preclude addressing them responsibly. As with scientific and medical experiments, we should conduct them responsibly, "minimizing negative and unwanted side effects to make the best of technologies that can greatly improve our lives" (Robaey, 2016, p. 899).

Van de Poel explicitly relies on Collingridge to discard the practice of anticipation as a way to foresee technological impacts. For him, anticipation runs "a risk of missing out on important actual social consequences of new technologies and of making us blind to surprises" (Van de Poel, 2016, p. 668). While he does acknowledge the value of scenarios for public engagement and deliberation, he questions their value for the responsible introduction of new technologies: scenarios direct the attention of the public away from real ethical issues and toward unlikely speculative futures (ibid., p. 670).

Conclusion

To support the responsible introduction of new technologies, Van de Poel (2011; 2016) provides an ethical framework for social experiments that comprises four bioethical principles. He further subdivides these principles into sixteen ethical initial conditions[4] (Van de Poel, 2011, p. 289) that can help experimenters to implement the principles in practice.

While the approach of sociotechnical experiments offers robust guidelines for responsible social experimentation and encourages forward-looking responsibility, it must be augmented with a method to "look forward" in a well-grounded manner. Any ambition to allow sociotechnical experiments to be more than trial-and-error requires a good, yet modest, instrument to look forward in a substantial manner.

Technological Mediation

The approach of technological mediation offers a third way to deal with the Collingridge dilemma in the ethics and governance of technology. Rather than discussing value change in the distant-future or conducting responsible social experiments with technology, it studies how technologies mediate value frameworks at existing and imaginary thresholds, thus exploring the foundation of technomoral change itself. As I have demonstrated in earlier chapters, with the help of the lemniscate principle of moral sense-making and the IPA method, the technological mediation approach allows the examination of how people implicitly articulate value conceptions and provide existing values with new meanings in relation to the anticipated mediating roles of technology.

As I have discussed in chapter 2, because of their common focus on the interaction between technology and morality, the approach of technological mediation possesses a close affinity with the technomoral change approach. However, while the approach of technomoral change assumes the connection between technological and moral developments and aims to anticipate future moral change in relation to technological innovations, the mediation approach enables to study actual processes of technomoral change in practice at the phenomenological micro-level, which ultimately forms the basis of macro-level technomoral developments. Because of its focus on the micro-level of human-technology relations, applying the approach of technological mediation allows the study of how moral frameworks develop in interaction with technological developments.

At the same time, the lemniscate principle of interpretation, coupled with the IPA method, lends the appropriation study a necessary hermeneutic sensibility. This explicates the hermeneutic dimension of value dynamism and change while both avoiding the trap of treating the research participants as repositories of values and accounting for the non-neutral role of the researcher in the setup and analysis of the study.

For instance, the mediation analysis of YouTube comments in chapter 3 demonstrates how moral hermeneutics accompanies the introduction of Glass. In particular, the study tentatively illustrates how the introduction of Glass might mediate the social practice of communication, the responsibility and proportionality of using Glass in public and private encounters, and the relation of Glass to memory making and to maintaining expectation of privacy in public places. The appropriation study suggests how people anticipate the mediating role of Glass in their daily experiences and practices and how, in connection to this, specific articulations of privacy become visible. The technological mediation approach does not provide generalizing predictions on the possible societal or ethical impacts of Google Glass, nor does it apply static moral conceptions to approach the device. Rather, it explores specific human practices and experiences to identify how the introduction of Glass might fit or conflict with them, thus enabling the rearticulation of ethical concerns.

My earlier discussions on participants as constructing existing and imaginary thresholds of appropriation might fuel suspicions that the technological mediation approach is itself speculative, even if it delegates speculation to research participants. I wish to preemptively dispel any such suspicions by noting that the mediation approach focuses not on distant-future predictions but on modest anticipations that are empirically and philosophically grounded in the current moral dynamics.

The appropriation study is interested in how people make sense of technologies within their experiences and sociocultural settings and how, amid this, certain directions of moral hermeneutics become available for reflection. This phenomenological task would be unachievable unless the participants themselves took charge of the discussion, beginning with their pre-judgments and unique histories. Even though participants inevitable rely on proactive agency to both project and reflect on the situations with emerging technologies, the goal is not to assess the anticipations that participants make or to take them as a given (which indeed would be speculative). Rather, the goal is to understand how these discussions reveal present-based moral hermeneutics, how certain value facets become visible, contextualized, and rearticulated, and how the accents between them shift already now. Such explorations of value dynamism in the present, coupled with studying existing technological practices, ethical debate, and technological visions, form the basis for modest anticipation about prospective value change.

Although some of these anticipated value mediations will not eventually manifest in reality while many other unforeseen ones will, identifying how and why potential avenues of moral hermeneutics may crystallize is important to understanding and accounting for the role of technology in mediating ethical concerns. As the IPA study in the case of sex selection has demonstrated, even an analysis based on a small study sample can draw a rich landscape of ethical concerns, thus re-articulating the existing ethical debate about sex selection and introducing new ethical challenges that are sensitive to the sociomaterial context. An IPA-based appropriation study can thus expand and deepen the ethical debate about technologies with considerations of moral hermeneutics while avoiding the charge of speculation.

A study of the dynamics of technological mediation and appropriation surrounding a technology, be it in the existing or foreseen practices, opens a new way of addressing the moral dimension of technology that provides one escape from the ethical variant of the Collingridge dilemma. To remind the reader, according to the value-articulated dilemma, at an early stage of development, we do not yet know how a technology will affect the value frameworks through which it will be evaluated in the future, while at a later stage, its implications for society and morality are clearer, but it is more difficult to guide the development in a certain direction. Complementing the approaches of technomoral change and sociotechnical experiments, the technological mediation approach demonstrates that an empirically informed method to anticipate the impact of technology on value frameworks, which contextualizes the future-oriented character of the technomoral scenario approach and moves beyond the rejection of anticipation by the sociotechnical experiments approach, does indeed exist.

One could argue that the mediation approach is in fact very close to the technomoral change approach, since both reveal how technologies can affect moral frameworks. The mediation approach extends beyond identifying future soft impacts; its focus on the mediating role of technologies in human-world relations enables it to develop detailed analyses of the present-based implications of technologies for the practices, perceptions, and frameworks of users. One could also argue that the Glass Explorers and YouTube commenters are in fact participating in (and even conducting) a sociotechnical experiment with little sense of direction and no guidance, transforming the societal and moral canvas along the way. Yet, drawing on Verbeek (2010), if we were to conduct this social experiment deliberately and sensibly, aiming to develop meaningful relations with such experimental technologies, we would also need to include well-informed anticipations of the ways in which technologies help to shape human existence and mediate moral frameworks. Throughout the appropriation studies in this book, I have demonstrated how the mediation approach makes this possible.

The technological mediation approach, then, offers a way to understand how people engage or foresee engagement with technologies: how technologies impact or could impact their daily lives, the concerns that come to the surface, and amid all of this, specific moral understandings are being invented and reinvented in interactions with new technologies. If we are to responsibly engage with new technologies, the technological mediation approach, guided by the lemniscate principle and accompanied by an appropriation study, could be part of the (moral) learning process. The mediation approach enables anticipation and critical reflection on how technologies mediate human practices, experiences, and value frameworks. Rather than being "too late"—able to see the implications but without room to change the social role of the technology—or "too early"—able to intervene but without having clarity about the societal implications, this approach seems to be positioned "just in time."

An empirically philosophical grounding of the technological mediation approach can potentially lend itself to settings where technologies are being discussed or experimented with just before they are introduced on a largescale. The mediation approach makes it possible to anticipate and reflect on the ethical implications of technology to ensure the informed design, use, and governance of the new technologies. The following, and concluding section of the book, reflects on this and other findings from the book and sketches several avenues for further explorations.

CONCLUSIONS

To conclude the book, I first briefly state what I have achieved in this project before I reflect upon the value of the findings for the larger fields of technology ethics and design discussed in this chapter.

In this book, I have introduced a developmental and relational account of values, sensitive to their sociomaterial embedding and open to change. This has been accomplished by emphasizing the pragmatist origins of the technological mediation approach, which highlights the interrelational ontology of values with their surrounding environment. This has allowed me to clarify the relation between technologies and values in what I called "moral hermeneutics," an account of moral sense-making as expanded with the active role of technologies and the sociocultural environment. The account of moral hermeneutics also expands the technological mediation approach, which lies at its basis, in that it transcends the idea of moral mediation introduced by Verbeek (2011), which suggests that technologies co-shape the moral perceptions, inclinations, and actions of people. I augment the technological mediation approach with the idea that technologies also mediate the meaning of values and that values can change in interaction with technologies. The result can be framed as a pragmatic and hermeneutic approach to how technologies mediate values.

In addition, I have developed a method for studying the dynamics of moral hermeneutics and the mediating role of technologies in it. Expanding upon

Verbeek (2015), I suggest that through the projective and practical appropriation of technologies, values resurface and become available for reflection and re-articulation. I have relied on Gadamer's principle of the hermeneutic circle (1975/2004) and connected it to the material hermeneutic ideas of Ihde (1998; 2005) to produce an encompassing principle of (moral) interpretation that accounts for the active role of people, technologies, and the world in the constitution of meaning, including the meaning of values. I have dubbed this principle the "hermeneutic lemniscate" and suggested that it can be used to trace and analyze moral hermeneutics within the fluid human-technology-world relations.

The hermeneutic lemniscate helps to clarify the moral sense-making and the mediation of technologies in this regard in several important ways. Firstly, the fluidity and interrelation that the model emphasizes preclude from giving the defining mediating role in morality to either human actors, specific technologies, or cultural elements. Therefore, even though technologies appear in the middle of the hermeneutic lemniscate, this does not mean that they are the constant defining mediator of the moral sense-making. Rather, the specific moral problems and the workable solutions to them, i.e., values, crystallize in the unique relation of the human, technological, and cultural counterparts. Secondly, neither of these counterparts can be the only source of values that produces the moral practice at hand. People have a certain moral compass that they intentionally or unconsciously translate into their everyday practices with technologies. As this book shows, technologies can also explicitly promote certain values that went into their design, for instance, the privacy-sensitive Corona-tracking technologies. Notwithstanding this, technologies may reflect additional values of the designers and of the larger cultural environment that indirectly gave shape to the design process. In the case of Corona-tracking technologies, their integration with smartphones communicates an assumption that everyone can have access to smartphones and can use them proficiently, whereas there are multiple groups in society who risk being excluded by a reliance on such apps during the pandemic due to the varying abilities, socioeconomic standing, etc. So the values of inclusion and diversity are also raised by the use of these apps that promote privacy, comfort, and ease of use. Additionally, one may argue that a big focus on privacy was a reflection of the Western moral push based on prior negative experiences with digital technologies and the companies producing them. However important, as the citizen participation studies demonstrate (Verbeek et al., 2020), overemphasizing privacy in the development of such apps tends to compromise other important values of the users, such as solidarity. The hermeneutic lemniscate allows navigating this and other moral practices from either the entry point of cultural systems, the technology at hand, or specific users. As such, it allows taking into account multiple value sources without favoring either one of them by default and exploring what may arise at their intersection.

Finally, in this book, I have developed a method to empirically study technologically mediated moral hermeneutics. To this end, I have introduced the method of Interpretative Phenomenological Analysis, or IPA, which originates from psychology and studies the attribution and construction of meaning (Smith, Flowers & Larkin, 2009), and narrowed its focus to experiences with technologies while expanding its scope to include an anticipative dimension. With this, IPA can empirically study the detailed workings of the lemniscate principle of interpretation and can unravel multiple possible or existing directions of moral hermeneutics. More specifically, the lemniscate model helps to clarify how, throughout the process of appropriation, different values surface, undergo rearticulation and make space for new value meanings. However, the idiographic focus of IPA on the micro-scale of individuals and their sociocultural lives requires manual approaches to collecting and analyzing the data, which makes the use of this method time- and effort-consuming, as well as presenting a small sample of respondents.

Nonetheless, the IPA method is very useful in clarifying the rich landscape of moral hermeneutics and making it available for reflection. The method allows for a richer form of mediation analysis compared to the comment-based experimental study in chapter 3. It also accounts for the hermeneutic level of values, currently unexposed in other approaches dealing with values and technologies (e.g., Swierstra et al., 2009; Van de Poel, 2015). Furthermore, IPA adds a constructive element to the technological mediation approach that could make it potentially applicable to the fields of design and policy-making. Lemniscate-guided, IPA-based mediation analysis does not aim to evaluate technologies from the outside or pass a judgment based on external criteria. Rather, such an analysis remains close to the field of technological development—the actual or anticipated practices with technology to accompany its informed use, design, and decision-making on a broader scale.

In this chapter, I have also provided several brief conclusions regarding the conceptual findings of this book. Specifically, I zoomed out from the empirical level to reflect on what the moral hermeneutics account that acknowledges the mediating role of technologies implies for the field of design and ethics. To this end, I have reintroduced the ethical variant of the Collingridge dilemma, briefly explained in the introduction. This variant builds on the technological mediation of values and suggests that we design and use technologies with certain values in mind, while these technologies simultaneously help to redefine the meaning of the values we use to develop and evaluate them. I have explained how the mediation approach allows the study of moral hermeneutics in relation to technologies in an empirically and philosophically grounded manner. I have also suggested that the expanded technological mediation account, as developed in this book, could facilitate a move beyond the dilemma and complement existing approaches addressing value change induced by technologies. Namely, with regard to the technomoral change approach (Swierstra, 2013), it can provide a solid empirical basis to understand the current value dynamism that can later crystallize in value change. Concerning the sociotechnical experimentation approach (Van de Poel, 2013), the mediation approach can introduce a method for a modest yet empirically grounded form of anticipation to complement the regulative efforts of sociotechnical experimentation. I do not outline how mediation as an escape from a dilemma actually functions in practice, for this largely exceeds the scope of the present research. All chapters until this point have been necessary to arrive at this dilemma and explain its significance. A practical study could be a useful next step to further develop the mediation approach and the account of moral hermeneutics in general.

Lastly, I have returned to the findings regarding the lemniscate principle of interpretation and the IPA-guided method to study moral hermeneutics in the field of technology ethics and design. More specifically, I have returned to the discussion of the Value Sensitive Design approach and suggested that even though an appropriation study carries several limitations, integrating it into the VSD approach could nonetheless be worthwhile. This could practically augment the VSD stages of conceptual and empirical investigations and conceptually expand upon and substantiate the philosophical understanding of values in VSD. Throughout this chapter, I have discussed the potential challenges of implementing an appropriation study, which are primarily related to its time-consuming nature and the potential need to involve a philosopher in some way.

One future avenue of research for the follow-up study of moral hermeneutics could be zooming in on the role of culture or the world in human-technology-world relations as co-productive for the value frameworks. As seen from the explorative appropriation study of Google Glass in chapter 3 and the IPA study of the sex selection chip in chapter 4, culture is arguably the broadest and most difficult to account for dimension in the study of moral sense-making. Therefore, it would be interesting to find specific ways to structurally and methodologically acknowledge its productive role in the moral hermeneutics. With this, I do not aim for the formalization of culture for the benefit of the naïve simplicity afforded to us by formulas or predefined pathways. Nonetheless, discerning patterns akin to the ways in which value change can proceed (van de Poel & Kudina, 2022), only with regard to how the sociocultural setting, alongside technologies, plays a role in morality, would augment the study of moral hermeneutics without pretending to exhaust either the cultural diversity or the pathways through which our norms, traditions, and institutional frameworks co-constitute our moral landscape. In addition to the qualitative empirical methods, here, it would be interesting once again to break the methodological horizons and to consider the modelling and simulation methods, such as Multi-Agent Systems Modelling or Choice Behavior Modelling (e.g., Van Dam et al., 2012), that could design complex interactions between different agents and potentially showcase how moral hermeneutics plays out in the interaction of the different components of the sociotechnical systems (e.g., de Wildt et al., 2020; 2021).

In this book, I have expanded the technological mediation account with considerations of value dynamism and change, and designed an appropriation study to examine moral hermeneutics in relation to existing and emerging technologies. I have also addressed the question of how we can practically deal with the idea of moral hermeneutics. The type of empirically informed philosophy that I have developed throughout the chapters offers one way to do so. As I have shown in this book, the mediation approach is equipped to study moral hermeneutics in the case of both technologies with prototypes and those existing predominantly in the minds of people. The appropriation study, guided by a lemniscate principle through an IPA method, offers the means to remain simultaneously engaged and distanced when studying how technologies mediate value frameworks. The appropriation study reveals that the values used to evaluate technologies are not independent from these technologies but rather are co-constituted by them. The appropriation study allows the exploration of the moral hermeneutics of both existing technologies and those at the threshold of introduction, from outside and within, both anticipatively and empirically. With the help of the technological mediation approach and an appropriation study, those engaging in ethics and the design of technology can construct empirically grounded situations to understand the moral hermeneutics involved. These findings can not only contribute to theoretical discussions in ethics but also possess the potential to facilitate ethical reflection about technologies in formal decision-making and contribute to the more responsible design and use of technologies.

NOTES

1. The comprehensive list of values endorsed by Value Sensitive Design is as follows: human welfare, ownership and property, privacy, freedom from bias, universal usability, trust, autonomy, informed consent, accountability, identity, calmness, and environmental sustainability (Friedman and Kahn, 2003).

2. A modified version of this section appeared in the following open-access article: Kudina, O., and P.-P. Verbeek. (2019). Ethics from within: Google Glass, the Collingridge dilemma, and the mediated value of privacy. *Science, Technology, & Human Values, 44*(2), pp. 291-314.

3. There is an ongoing work by some researchers within the technomoral change approach to explicate how specific technology induced moral destabilizations crystalize in the present (see Weingartz, S., dissertation in progress). Nonetheless, to my best knowledge, it does not address the issue I try to engage with in this sub-section, namely how values which we use to design and evaluate technologies change in relation to these same technologies and how to account for that.

4. The sixteen (originally thirteen (Van de Poel, 2011, p. 289)) initial ethical conditions are: (1) Absence of other reasonable means for gaining knowledge about risks and benefits; (2) Monitoring of data and risks while addressing privacy concerns; (3) Possibility and willingness to adapt or stop the experiment; (4) Containment of risks as far as reasonably possible; (5) Consciously scaling up to avoid large-scale harm and to improve learning; (6) Flexible set-up of the experiment and avoidance of lock-in of the technology; (7) Avoid experiments that undermine resilience; (8) Reasonable to expect social benefits from the experiment; (9) Clear distribution of responsibilities for setting up, carrying out, monitoring, evaluating, adapting, and stopping of the experiment; (10) Experimental subjects are informed; (11) The experiment is approved by democratically legitimized bodies; (12) Experimental subjects can influence the setting up, carrying out, monitoring, evaluating, adapting, and stopping of the experiment; (13) Experimental subjects can withdraw from the experiment; 13. Vulnerable experimental subjects are either not subject to the experiment or are additionally protected or particularly profit from the experimental technology (or a combination); 15. A fair distribution of potential hazards and benefits; 16. Reversibility of harm or, if impossible, compensation of harm (Van de Poel, 2016).

References

Aagaard, J. (2018). Entering the portal: Media technologies and experiential transportation. In J. Aagard, J. K. Berg Friis, J. Sorenson, O. Tafdrup, & C. Hasse (Eds.), *Postphenomenological methodologies: New ways in mediating techno-human relationships* (pp. 45–62). Lanham: Lexington Books.

Aagaard, J., Berg Friis, J. K., Sorenson, J., Tafdrup, O., & Hasse, C. (Eds.). (2018). *Postphenomenological methodologies: New ways in mediating techno-human relationships*. Lanham: Lexington Books.

Alsheikh, T., Rode, J. A., & Lindley, S. E. (2011). (Whose) value-sensitive design: A study of long-distance relationships in an Arabic cultural context. Paper presented at the *Proceedings of the ACM 2011 Conference on Computer Supported Cooperative Work*, March 19–23, Hangzhou, China.

Baruch, J. D., Kaufman, D., & Hudson, K. L. (2008). Genetic testing of embryos: Practices and perspectives of US In Vitro Fertilization clinics. *Fertility and Sterility, 89*(5), 1053–1058.

Bayefsky, M., & Jennings, B. (2015). *Regulating preimplantation genetic diagnosis in the United States: The limits of unlimited selection.* New York: Palgrave Macmillan US.

Berendsen, J. T., Kruit, S. A., Atak, N., Willink, E., & Segerink, L. I. (2020). Flow-free microfluidic device for quantifying chemotaxis in spermatozoa. *Analytical Chemistry, 92*(4), 3302–3306.

Berker, T., Hartmann, M., Punie, Y., & Ward, K. (2006b). Introduction. In T. Berker, M. Hartmann, Y. Punie, & K. Ward (Eds.), *Domestication of media and technology* (pp. 1–17). Berkshire: Open University Press.

Berker, T., Hartmann, M., Punie, Y., & Ward, K. (Eds.). (2006a). *Domestication of media and technology*. Berkshire: Open University Press.

Bertel, T. F. (2018). Domesticating smartphones. In J. Vincent & L. Haddon (Eds.), *Smartphone cultures* (Chapter 7, pp. 83–94). London: Routledge.

Besmer, K. M. (2015). What robotic re-embodiment reveals about virtual re-embodiment: A note on the extension thesis. In R. Rosenberger & P.-P. Verbeek (Eds.),

Postphenomenological investigations: Essays on human–technology relations (pp. 55–72). Lanham: Lexington Books.

Blank, G., Bolsover, G., & Dubois, E. (2014). A new privacy paradox: Young people and privacy on social network sites. Paper presented at the *Annual Meeting of the American Sociological Association,* San Fransisco, California, August 17.

Blond, L., & Schiølin, K. (2018). Lost in translation?: Getting to grips with multistable technology in an apparently stable world. In J. Aagard, J. K. Berg Friis, J. Sorenson, O. Tafdrup, & C. Hasse (Eds.), *Postphenomenological methodologies: New ways in mediating techno-human relationships* (pp. 151–168). Lanham: Lexington Books.

Blyth, E., Frith, L., & Crawshaw, M. (2008). Ethical objections to sex selection for non-medical reasons. *Reproductive Biomedicine Online, 16,* 41–45.

Boenink, M., & Kudina, O. (2020). Values in responsible research and innovation: From entities to practices. *Journal of Responsible Innovation, 7*(3), 450–470.

Boenink, M., & Swierstra, T. (2015). Technomoral scenarios. Paper presented at the workshop *What's Next in Socio-Technical Intervention Approaches?* University of Twente, Enschede, the Netherlands, June 22–23.

Boenink, M., Swierstra, T., & Stemerding, D. (2010). Anticipating the interaction between technology and morality: A scenario study of experimenting with humans in bionanotechnology. *Studies in Ethics, Law, and Technology, 4*(2), Article 4. https://doi.org/10.2202/1941-6008.1098.

Bonfert, M., Spliethöver, M., Arzaroli, R., Lange, M., Hanci, M., & Porzel, R. (2018). If you ask nicely: A digital assistant rebuking impolite voice commands. In *Proceedings of the 20th ACM International Conference on Multimodal Interaction* (pp. 95–102). New York: Association for Computing Machinery.

Borgmann, A. (1984). *Technology and the character of contemporary life: A philosophical inquiry* (Vol. 20). Chicago: University of Chicago Press.

Borgmann, A. (1999). *Holding on to reality: The nature of information at the turn of the millennium.* Chicago and London: University of Chicago Press.

Borning, A., & Muller, M. (2012). Next steps for value sensitive design. In *Proceedings of the SIGCHI Conference on Human Factors in Computing Systems* (pp. 1125–1134). New York: ACM.

Brey, P. (2000). Method in computer ethics: Towards a multi-level interdisciplinary approach. *Ethics and Information Technology, 2*(2), 125–129.

Bugeja, J., Jacobsson, A., & Davidsson, P. (2016). On privacy and security challenges in smart connected homes. *2016 European Intelligence and Security Informatics Conference* (pp. 172–175). IEEE.

Burton, N. G., & Gaskin, J. (2019). "Thank you, Siri": Politeness and intelligent digital assistants. In *The Americas Conference on Information Systems 2019 Proceedings* (pp. 1–10). Atlanta: AIS.

Calo, M. R., Friedman, B., Kohno, T., Almeter, H., & Logler, N. (Eds.). (2020). *Telling stories: On culturally responsive artificial intelligence: 19 stories.* University of Washington Tech Policy Lab.

Camp, L. J. (2003). Designing for trust. In R. Falcone, S. Barber, L. Korba, & M. Singh (Eds.), *Trust, reputation, and security: Theories and practice: AAMAS 2002 international workshop*, Bologna, Italy, July 15, 2002. Selected and Invited Papers (pp. 15–29). Berlin, Heidelberg: Springer.

Capurro, R. (2010). Digital hermeneutics: An outline. *AI & Society, 25*(1), 35–42.

Carter, S., Green, J., & Thorogood, N. (2013). The domestication of an everyday health technology: A case study of electric toothbrushes. *Social Theory & Health, 11*(4), 344–367.

Charmaz, K. (2006). *Constructing grounded theory: A practical guide through qualitative analysis.* London: Sage.

Chenail, R. J. (2011). YouTube as a qualitative research asset: Reviewing user generated videos as learning resources. *The Qualitative Report, 16*(1), 229–235.

Coeckelbergh, M. (2012). *Growing moral relations: Critique of moral status ascription.* New York: Palgrave Macmillan.

Collingridge, D. (1980). *The social control of technology.* New York: St. Martin's Press.

Cummings, M. L. (2006). Integrating ethics in design through the value-sensitive design approach. *Science and Engineering Ethics, 12*(4), 701–715.

De Boer, B., Te Molder, H., & Verbeek, P. P. (2020). Constituting 'visual attention': On the mediating role of brain stimulation and brain imaging technologies in neuroscientific practice. *Science as Culture, 29*(4), 503–523.

De la Bellacasa, M. P. (2011). Matters of care in technoscience: Assembling neglected things. *Social Studies of Science, 41*(1), 85–106.

De Wagenaar, B. et al. (2015). Microfluidic single sperm sntrapment and analysis. *Lab on a Chip, 15*(5), 1294–1301.

de Wildt, T. E., Chappin, E. J. L., van de Kaa, G., Herder, P. M., & van de Poel, I. R. (2020). Conflicted by decarbonisation: Five types of conflict at the nexus of capabilities and decentralised energy systems identified with an agent-based model. *Energy Research & Social Science, 64,* 101451.

de Wildt, T. E., van de Poel, I. R., & Chappin, E. J. L. (2021). Tracing long-term value change in (energy) technologies: Opportunities of probabilistic topic models using large data sets. *Science, Technology, & Human Values.* https://doi.org/10.1177/01622439211054439.

Dewey, J. (1922). *Human nature and conduct. An introduction to social psychology.* New York: H. Holt and Company.

Dewey, J. (1929). *Experience and nature.* London: George Allen & Unwin, Ltd.

Dewey, J. (1930). *The quest for certainty.* London: George Allen & Unwin, Ltd.

Dewey, J. (1939). *Theory of valuation.* Chicago: University of Chicago Press.

Dewey, J. (1940). Time and Individuality. In D. W. Hering, W. F. G. Swann, J. Dewey, & A. H. Compton (Eds.), *Time and its mysteries, series II* (pp. 85–112). New York: New York University Press (Original work published in 1938).

Dewey, J. (1976a). Creative democracy: The task before us. In J. Boydston (Ed.), *John Dewey: The later works, 1925–1953* (Vol. 14, pp. 224–230). Carbondale: Southern Illinois University Press (Original work published 1939).

Dewey, J. (1976b). In J. Gouinlock (Ed.), *The moral writings of John Dewey.* New York: Hafner Press.

Dewey, J., & J. H. Tufts. (1932). *Ethics* (Rev. ed.). New York: Henry Holt and Company (Original work published 1908).

Dickens, B. (2002). Can sex selection be ethically tolerated? *Journal of Medical Ethics, 28*(6), 335–336.

Dodd, D. (2020, April 20). Contact-tracing apps raise surveillance fears. *Financial Times*. Accessed on December 24, 2021 from https://www.ft.com/content/005ab1a8-1691-4e7b-8e10-0d3d2614a276.

Dorrestijn, S. (2012). *The design of our own lives: Technical mediation and subjectivation after Foucault* [PhD thesis]. Enschede: University of Twente.

Dreyfus, H. L. (1972). *What computers can't do: A critique of artificial reason*. New York: Harper & Row.

Dreyfus, H. L. (2001). *On the Internet*. New York: Routledge.

Druga, S., Williams, R., Breazeal, C., & Resnick, M. (2017). "Hey Google is it OK if I eat you?" Initial explorations in child-agent interaction. In *Proceedings of the 2017 Conference on Interaction Design and Children* (pp. 595–600). New York: Association for Computing Machinery.

Dussauge, I., Helgesson, C. F., Lee, F., & Woolgar, S. (2015). On the omnipresence, diversity, and elusiveness of values in the life sciences and medicine. In I. Dussauge, C.-F. Helgesson, & F. Lee (Eds.), *Value practices in the life sciences and medicine* (pp. 1–30). Oxford: Oxford University Press.

Edwards, D., & Potter, J. (1992). *Discursive psychology*. London: Sage.

European Parliament. (2002). *Directive 2002/58/EC of the European Parliament and of the Council of 12 July 2002 concerning the processing of personal data and the protection of privacy in the electronic communications sector*. European Union: European Parliament, Council of the European Union.

Fairclough, N. (2013). *Critical discourse analysis: The critical study of language*. Routledge.

Fessler, L. (2017, February 22). Siri, define patriarchy: We tested bots like Siri and Alexa to see who would stand up to sexual harassment. *Quartz*. Accessed on April 4, 2022 from https://qz.com/911681/we-tested-apples-siri-amazon-echos-alexa-microsofts-cortana-and-googles-google-home-to-see-which-personal-assistant-bots-stand-up-for-themselves-in-the-face-of-sexual-harassment/.

Firstenberg, A., & Salas, J. (2014). *Designing and developing for Google Glass*. Sebastopol: O'Reilly Media, Inc.

Flanagan, M., Howe, D. C., & Nissenbaum, H. (2005). Values at play: Design tradeoffs in socially-oriented game design. In *Proceedings of the SIGCHI Conference on Human Factors in Computing Systems* (pp. 751–760). New York: ACM.

Frankena, W. K. (1967). Value and valuation. In P. Edwards (Ed.) *The Encyclopedia of philosophy* (pp. 636–641). New York : Macmillan.

Friedman, B., & Hendry, D. G. (2019). *Value sensitive design: Shaping technology with moral imagination*. MIT Press.

Friedman, B., & Kahn, P. (2003). Human values, ethics, and design. In J. A. Jacko & A. Sears (Eds.), *The human-computer interaction handbook* (pp. 1177–1201). L. Erlbaum Associates Inc.

Friedman, B., Kahn, P., & Borning, A. (2002). *Value sensitive design: Theory and methods* (pp. 02–12). University of Washington Technical Report.

Friedman, B., Kahn, P., & Borning, A. (2006). Value sensitive design and information systems. In P. Zhang & D. F. Galletta (Eds.), *Human-computer interaction and management information systems* (pp. 348–373). Armonk: M. E. Sharpe.

Friedman, B. (1999). *Value-Sensitive Design: A research agenda for information technology*. Contract No: SBR-9729633. Arlington: National Science Foundation.

Fry, P. (2009). *Ways in and out of the hermeneutic circle* [Open Yale courses]. University of Yale. Accessed on April 4, 2022 from http://oyc.yale.edu/english/engl-300/lecture-3.

Gadamer, H.-G. (1977). *Philosophical hermeneutics* (D. E. Linge, Trans.). Berkeley: University of California Press.

Gadamer, H.-G. (2004). *Truth and method* (J. Weinsheimer & D. G. Marshall, Trans., 2nd ed.). New York: Crossroad (Original work published 1975).

Gadamer, H.-G. (2006). Language and understanding (1970). *Theory, Culture & Society, 23*(1), 13–27.

Genderlessvoice.com. (2019). *Meet Q, the first genderless voice* [official website]. Accessed on April 4, 2022 from https://www.genderlessvoice.com/about.

GenderSelect. (2017). *Methods of gender selection*. Accessed on March 29, 2022 from http://chooseagender.com/Methods-Of-Gender-Selection.aspx.

Google. (2013). Glass security & privacy. *Google Glass website*. Accessed on April 5, 2022 from https://sites.google.com/site/glasscomms/faqs#GlassSecurity&Privacy.

Google. (2014). Explorers: Do's and don'ts. *Google Glass website*. Accessed on April 5, 2022 from https://sites.google.com/site/glasscomms/glass-explorers.

Google. (2015). Wink. *Google Glass website*. Accessed on April 5, 2022 from https://support.google.com/glass/answer/4347178?hl=en.

Grunwald, A. (2009). Technology assessment: Concepts and methods. In A. Meijers (Ed.), *Philosophy of technology and engineering sciences* (pp. 1103–1146). Amsterdam: North Holland.

Grunwald, A. (2016). *The hermeneutic side of responsible research and innovation*. John Wiley & Sons.

Grunwald, A. (2020). The objects of technology assessment. Hermeneutic extension of consequentialist reasoning. *Journal of Responsible Innovation, 7*(1), 96–112.

Haen, D. (2015). *The politics of good food: Why food engineers and citizen-consumers are talking at cross-purposes* (PhD diss.). Maastricht: University of Maastricht.

Haen, D., Sneijder, P., te Molder, H., & Swierstra, T. (2015). Natural Food: Organizing 'responsiveness' in responsible innovation of food technology. In B.-J. Koops, I. Oosterlaken, H. Romijn, T. Swierstra, & J. van den Hoven (Eds.), *Responsible innovation 2* (pp. 161–181). Cham: Springer.

Halpern, B. M., van Son, R., Brekel, M. V. D., & Scharenborg, O. (2020). Detecting and analysing spontaneous oral cancer speech in the wild. *Interspeech Confrence 2020*. Preprint: arXiv:2007.14205.

Hämäläinen, N. (2016). *Descriptive ethics: What does moral philosophy know about morality?* New York: Palgrave Macmillan.

Harper, J. C., & SenGupta, S. B. (2012). Preimplantation genetic diagnosis: State of the ART 2011. *Human Genetics, 131*(2), 175–186.

Hasse, C. (2008). Postphenomenology: Learning cultural perception in science. *Human Studies, 31*(1), 43–61.

Hasse, C. (2018). Studying the telescopes of others: Toward a postphenomenological methodology of participant observation. In J. Aagard, J. K. Berg Friis, J. Sorenson, O. Tafdrup, & C. Hasse (Eds.), *Postphenomenological methodologies: New ways in mediating techno-human relationships* (pp. 241–258). Lanham: Lexington Books.

Heidegger, M. (1962). *Being and time* (J. Macquarrie & E. Robinson, Trans.). New York: Harper & Row (Original work published 1927).

Heritage, J., & Raymond, G. (2005). The terms of agreement: Indexing epistemic authority and subordination in talk-in-interaction. *Social Psychology Quarterly, 68*(1), 15–38.

Hewson, C., & Buchanan, T. (Eds.). (2013). *Ethics guidelines for Internet-mediated research*. Leicester: The British Psychological Society.

Honan, M. (2013, June 3). I, Glasshole: My Year With Google Glass. *Wired*. Accessed on April 5, 2022 from https://www.wired.com/2013/12/glasshole.

Hutchby, I., & Wooffitt, R. (2008). *Conversation analysis* (2nd ed.). Cambridge: Polity Press.

Ihde, D. (1979). *Technics and praxis. A philosophy of technology*. Boston: D. Reidel Publishing Company.

Ihde, D. (1990). *Technology and the lifeworld: From garden to earth*. Bloomington: Indiana University Press.

Ihde, D. (1993). *Philosophy of technology: An introduction*. New York: Paragon House.

Ihde, D. (1998). *Expanding hermeneutics: Visualism in science*. Evanston: Northwestern University Press.

Ihde, D. (2002). *Bodies in technology*. Minneapolis: University of Minnesota Press.

Ihde, D. (2005). Material hermeneutics. Paper presented in *The Meeting on Symmetrical Archaeology*. Stanford University, Theoretical Archaeology Group. Accessed on April 5, 2022 from https://web.archive.org/web/20070607044901/http://traumwerk.stanford.edu:3455/symmetry/746.

Ihde, D. (2008). *Ironic technics*. Automatic Press/VIP.

Ihde, D. (2009). *Postphenomenology and technoscience: The Peking university lectures*. New Tork: Suny Press.

Irwin, S. (2018). Affective algorithms: Vocal emotion in digital form. Paper presented at the *Conference Philosophy of Human-Technology Relations*, University of Twente, Enschede, the Netherlands, July 11–13.

Jacks, A., Haley, K. L., Bishop, G., & Harmon, T. G. (2019). Automated speech recognition in adult stroke survivors: Comparing human and computer transcriptions. *Folia Phoniatrica et Logopaedica, 71*(5–6), 282–292.

Jansen-Kosterink, S. M., Hurmuz, M., den Ouden, M., & van Velsen, L. (2021). Predictors to use mobile apps for monitoring COVID-19 symptoms and contact tracing: A survey among Dutch citizens. *JMIR Formative Research, 5*(12), e28416.

Jones, M. L. (2018). *Ctrl+ Z: The right to be forgotten*. NYU Press.

Khodamoradi, M., Rafizadeh Tafti, S., Mousavi Shaegh, S. A., Aflatoonian, B., Azimzadeh, M., & Khashayar, P. (2021). Recent microfluidic innovations for sperm sorting. *Chemosensors, 9*(6), 126.

Kiran, A. H., Oudshoorn, N., & Verbeek, P.-P. (2015). Beyond checklists: Toward an ethical-constructive technology assessment. *Journal of Responsible Innovation, 2*(1), 5–19.

Koelle, M., Kranz, M., & Möller, A. (2015). Don't look at me that way!: Understanding user attitudes towards data glasses usage. In *Proceedings of the 17th International Conference on Human-Computer Interaction with Mobile Devices and Services* (pp. 362–372). New York: ACM.

Kudina, O. (2019). Accounting for the moral significance of technology: Revisiting the case of non-medical sex selection. *Journal of Bioethical Inquiry, 16*(1), 75–85.

Kudina, O. (2021a). Bridging privacy and solidarity in COVID-19 contact-tracing apps through the sociotechnical systems perspective. *Glimpse, 22*(2), 43–54.

Kudina, O. (2021b). "Alexa, who am I?": Voice assistants and hermeneutic lemniscate as the technologically mediated sense-making. *Human Studies, 44*(2), 233–253.

Kudina, O., & Verbeek, P.-P. (2019). Ethics from within: Google Glass, the Collingridge dilemma, and the mediated value of privacy. *Science, Technology, & Human Values, 44*(2), 291–314.

Latour, B. (2004). Why has critique run out of steam? From matters of fact to matters of concern. *Critical Inquiry, 30*(2), 225–248.

Latour, B. (2005). *Reassembling the social: An introduction to actor—network theory*. Oxford: Oxford University Press.

Latour, B. (2008). *What is the style of matters of concern?* Assen: Royal Van Gorcum.

Lau, J., Zimmerman, B., & Schaub, F. (2018). Alexa, are you listening? Privacy perceptions, concerns and privacy-seeking behaviors with smart speakers. In *Proceedings of the ACM on Human-Computer Interaction* (Article 102, pp. 1–31). https://doi.org/10.1145/3274371.

Le Dantec, C. A., Poole, E. S., & Wyche, S. P. (2009). Values as lived experience: Evolving value sensitive design in support of value discovery. In *Proceedings of the SIGCHI Conference on Human Factors in Computing Systems* (pp. 1141–1150). New York: ACM.

Levy, S. (2017, July 18). Google Glass 2.0 is a startling second act. *Wired*. Accessed on April 5, 2022 from https://www.wired.com/story/google-glass-2-is-here.

Lim, S. S. (2008). Technology domestication in the Asian homestead: Comparing the experiences of middle class families in China and South Korea. *East Asian Science, Technology and Society: An International Journal, 2*(2), 189–209.

Lovett, L. (2020, 23 July). WHO's Director for Europe urges solidarity in using digital tools to combat Covid-19. *MobiHealthNews*. Accessed on December 24, 2021 from https://www.mobihealthnews.com/news/europe/whos-europe-director-urges-solidarity-using-digital-tools-combat-covid-19-0.

Lucivero, F. (2012). *Too good to be true. Appraising expectations for ethical technology assessment* (PhD thesis). Enschede: University of Twente.

Magnani, L. (2007). *Morality in a technological world: Knowledge as duty.* Cambridge: Cambridge University Press.

Manders-Huits, N. (2011). What values in design? The challenge of incorporating moral values into design. *Science and Engineering Ethics, 17*(2), 271–287.

Markham, A., & Buchanan, E. (Eds.). (2012). Ethical decision-making and Internet research: Recommendations from the AoIR Ethics Working Committee (Version 2.0). *Association of Internet Research.* Accessed on April 5, 2022 from http://aoir.org/reports/ethics2.pdf.

Mashable. (2013). Google Glass: Don't be a Glasshole. *YouTube.com.* Accessed on April 5, 2022 from https://www.youtube.com/watch?v=FlfZ9FNC99k.

Mayer-Schönberger, V. (2009). *Delete: The virtue of forgetting in the digital age.* Princeton: Princeton University Press.

MESA+. (2017). *Biomedical Microdevices: Spermatozoa On-Chip* [Project description]. Accessed on March 29, 2022 from https://www.utwente.nl/en/eemcs/bios/research/biomedical/Sperm_onchip.pdf.

Microsoft Corporation. (2015). HoloLens. *HoloLens Website.* Accessed on April 5, 2022 from http://www.microsoft.com/microsoft-hololens/en-us.

Mol, A. (1997). *Wat is kiezen? Een empirisch-filosofische verkenning* [What is choosing? An empirical-philosophical exploration]. (Inaugural Lecture). Enschede: University of Twente.

Mol, A. (2002). *The body multiple: Ontology in medical practice.* Durham and London: Duke University Press.

Mol, A. (2010). Actor-network theory: Sensitive terms and enduring tensions. *Kölner Zeitschrift für Soziologie und Sozialpsychologie, 50*(1), 253–269.

Myers, G. (2004). *Matters of opinion: Talking about public issues.* Cambridge: Cambridge University Press.

Nass, C. I., & Brave, S. (2005). *Wired for speech: How voice activates and advances the human-computer relationship.* Cambridge, MA: MIT press.

Nass, C., Moon, Y., & Green, N. (1997). Are machines gender neutral? Gender-stereotypic responses to computers with voices. *Journal of Applied Social Psychology, 27*(10), 864–876.

Office of the Privacy Commissioner of Canada. (2013). *Data protection authorities urge Google to address Google Glass concerns.* [News release]. Accessed on April 5, 2022 from https://www.priv.gc.ca/en/opc-news/news-and-announcements/2013/nr-c_130618/.

Orlikowski, W. J. (2007). Sociomaterial practices: Exploring technology at work. *Organization studies, 28*(9), 1435–1448.

Palanica, A., Thommandram, A., Lee, A., Li, M., & Fossat, Y. (2019). Do you understand the words that are comin outta my mouth? Voice assistant comprehension of medication names. *NPJ Digital Medicine, 2*(1), 1–6.

Parens, E. (2015). *Shaping our selves: On technology, flourishing, and a habit of thinking.* New York: Oxford University Press.

Parliamentary Office of Science and Technology. (2016). *POSTnote no. 198: Sex selection.* Accessed on April 5, 2022 from www.parliament.uk/documents/post/pn198.pdf.

Pitt, J. C. (2014). "Guns don't kill, people kill"; Values in and/or around technologies. In P. Kroes & P.-P. Verbeek (Eds.), *The moral status of technical artefacts* (pp. 89–101). Dordrecht: Springer.

Potter, J. (1996). *Representing reality: Discourse, rhetoric and social construction.* London: Sage.

Potts, A. (2015). "LOVE YOU GUYS (NO HOMO)." How gamers and fans play with sexuality, gender, and Minecraft on YouTube. *Critical Discourse Studies, 12*(2), 163–186.

Project Alias. (2018). Official website. Accessed on April 4, 2022 from https://bjoernkarmann.dk/project_alias.

Pyae, A., & Scifleet, P. (2018). Investigating differences between native English and non-native English speakers in interacting with a voice user interface: A case of Google Home. In *Proceedings of the 30th Australian Conference on Computer-Human Interaction* (pp. 548–553). New York: Association for Computing Machinery.

Rathenau Institute. (March 1996). *Bericht aan het Parlement: Meeste Nederlanders tegen geslachtskeuze.* [Letter to the Parliament: Majority of the Dutch people against the sex selection]. Den Haag: Rathenau Instituut.

Reckwitz, A. (2002). Toward a theory of social practices: A development in culturalist theorizing. *European Journal of Social Theory, 5*(2), 243–263.

Regan, P. (2002). Privacy as a common good in the digital world. *Information, Communication & Society, 5*(3), 382–405.

Reid, K., Flowers, P., & Larkin, M. (2005). Exploring lived experience: An introduction to interpretative phenomenological analysis. *The Psychologist, 18*(1), 20–23.

Reijers, W., & Coeckelbergh, M. (2020). *Narrative and technology ethics.* Cham: Palgrave Macmillan.

Reijers, W., Romele, A., & Coeckelbergh, M. (Eds.). (2021). *Interpreting technology: Ricoeur on questions concerning ethics and philosophy of technology.* Rowman & Littlefield.

Richardson, H. J. (2009). A 'smart house' is not a home: The domestication of ICTs. *Information Systems Frontiers, 11*(5), 599–608.

Rip, A., Misa, T. J., & Schot, J. (Eds). (1995). *Managing technology in society.* London: Pinter.

Robaey, Z. (2016). Gone with the wind: Conceiving of moral responsibility in the case of GMO contamination. *Science and Engineering Ethics, 22*(3), 889–906.

Roessler, B., & Mokrosinska, D. (2013). Privacy and social interaction. *Philosophy & Social Criticism, 39*(8), 771–791.

Rosenberger, R. (2009). The sudden experience of the computer. *AI & Society, 24*(2), 173–180.

Rosenberger, R. (2014). Multistability and the agency of mundane artifacts: From speed bumps to subway benches. *Human Studies, 37*(3), 369–392.

Rosenberger, R. (2017). *Callous objects: Designs against the homeless.* Minneapolis: University of Minnesota Press.

Rosenberger, R., & Verbeek, P.-P. (2015a). A field guide to postphenomenology. In R. Rosenberger & P.-P. Verbeek (Eds.), *Postphenomenological investigations: Essays on human–technology relations* (pp. 7–42). Lanham: Lexington Books.

Rosenberger, R., & Verbeek, P.-P. (Eds.). (2015b). *Postphenomenological investigations: Essays on human–technology relations*. Lanham: Lexington Books.
Sacks, H. (1992). *Lectures on conversation*. Oxford: Blackwell.
Sandel, M. (2004). The case against perfection. *The Atlantic Monthly, 293*(3), 51–62.
Savulescu, J. (1999). Sex selection: The case for. *The Medical Journal of Australia, 171*(7), 373–375.
Schuijff, M., & Munnichas, G. (Eds.). (2012). *Goed, beter, betwist. Publieksonderzoek naar mensverbetering* [Good, better, controversial. Public research on human enhancement.] [Rathenau Institute, Report]. Den Haag: Rathenau Instituut.
Schuster, M., Maier, A., Haderlein, T., Nkenke, E., Wohlleben, U., Rosanowski, F., ... Nöth, E. (2006). Evaluation of speech intelligibility for children with cleft lip and palate by means of automatic speech recognition. *International Journal of Pediatric Otorhinolaryngology, 70*(10), 1741–1747.
Schwandt, A. (2007). Idiographic interpretation. In A. Schwandt (Ed.), *The SAGE dictionary of qualitative inquiry* (3rd ed., pp. 145–146). Thousand Oaks, CA: Sage.
Scott, J., & Marshall, G. (2015). Value. In *Oxford Dictionary of Sociology* [Online edition]. Accessed on December 22, 2021 from https://bit.ly/2w5jaOa.
Secomandi, F. (2018). Service interfaces in human-technology relations: A case study of self-tracking technologies. In J. Aagard, J. K. Berg Friis, J. Sorenson, O. Tafdrup, & C. Hasse (Eds.), *Postphenomenological methodologies: New ways in mediating techno-human relationships* (pp. 86–102). Lanham: Lexington Books.
Segerink, L. I. et al. (2012). Microfluidic chips for semen analysis. *Journal of the International Federation of Clinical Chemistry and Laboratory Medicine, 23*(3), 66–69.
Shapiro, S. (1998). Places and spaces: The historical interaction of technology, home, and privacy. *The Information Society, 14*(4), 275–284.
Sharon, T. (2021). Blind-sided by privacy? Digital contact tracing, the Apple/Google API and big tech's newfound role as global health policy makers. *Ethics and Information Technology, 23*(1), 45–57.
Siffels, L. E. (2021). Beyond privacy vs. health: A justification analysis of the contact-tracing apps debate in the Netherlands. *Ethics and Information Technology, 23*(1), 99–103.
Silverstone, R., Hirsch, E., & Morley, D. (1994). Information and communication technologies and the moral economy of the household. In R. Silverstone & E. Hirsch (Eds.), *Consuming technologies; Media and information in domestic spaces* (pp. 13–28). Routledge: London.
Sintumuang, K. (2013, May 3). Google Glass: An Etiquette Guide. *Wall Street Journal*. Accessed on April 5, 2022 from https://on.wsj.com/2x7JyVM.
Smith, J. A. (2011a). Evaluating the contribution of interpretative phenomenological analysis. *Health Psychology Review, 5*(1), 9–27.
Smith, J. A. (2011b). Evaluating the contribution of interpretative phenomenological analysis: A reply to the commentaries and further development of criteria. *Health Psychology Review, 5*(1), 55–61.
Smith, J. A., Flowers, P., & Larkin, M. (2009). *Interpretative phenomenological analysis: Theory, method and research*. London: Sage.

Solove, D. J. (2002). Conceptualizing privacy. *California Law Review, 90*(4), 1087–1155.
Sorensen, K. (2006). Domestication: The enactment of technology. In T. Berker, M. Hartmann, Y. Punie, & K. Ward (Eds.), *Domestication of media and technology* (pp. 40–61). Berkshire: Open University Press.
Sowański, M., & Janicki, A. (2020). Leyzer: A dataset for multilingual Virtual Assistants. In P. Sojka, I. Kopecek, K. Pala, & A. Horak (Eds.), *International conference on text, speech, and dialogue proceedings* (pp. 477–486). Cham: Springer.
Spectacles. (2017). *Spectacles website*. Accessed on April 5, 2022 from https://www.spectacles.com.
Steeves, V., & Regan, P. (2014). Young people online and the social value of privacy. *Journal of Information, Communication and Ethics in Society, 12*(4), 298–313.
Suchman, L. (2007). *Human-machine reconfigurations: Plans and situated actions*. Cambridge: Cambridge University Press.
Swierstra, T,, & Rip, A. (2007). Nano-ethics as NEST-ethics: Patterns of moral argumentation about New and Emerging Science and Technology. *NanoEthics, 1*(1), 3–20.
Swierstra, T. (2011). *Heracliteïsche ethiek: omgaan met de soft impacts van technologie*. [Heraclitus' ethics: Dealing with the Soft Impacts of Technology][Inaugural lecture]. Maastricht: Maastricht University.
Swierstra, T. (2013). Nanotechnology and technomoral change. *Ethica & Politica/Ethics & Politics, 15*(1), 200–219.
Swierstra, T. (2016). Identifying technologically induced moral change. Paper presented at *the symposium Technology and morality*, University of Twente, Enschede, the Netherlands, November 10, 2016.
Swierstra, T., Stemerding, D., & Boenink, M. (2009). Exploring techno-moral change: The case of the obesitypill. In P. Sollie & M. Duwell (Eds.), *Evaluating new technologies* (pp. 119–138). Dordrecht: Springer.
Taylor, N. (2002). State surveillance and the right to privacy. *Surveillance & Society, 1*(1), 66–85.
Te Molder, H. (2008). Discursive psychology. In W. Donsbach (Ed.), *The International encyclopedia of communication* (pp. 1370–1372). John Wiley & Sons, Ltd.
Te Molder, H., & Potter, J. (Eds.). (2005). *Conversation and cognition*. Cambridge: Cambridge University Press.
Timmermans, J., Zhao, Y., & van den Hoven, J. (2011). Ethics and nanopharmacy: Value sensitive design of new drugs. *NanoEthics, 5*(3), 269–283.
Tonkiss, F. (2003). The ethics of indifference: Community and solitude in the city. *International Journal of Cultural Studies, 6*(3), 297–311.
Turkle, S. (Ed.). (2007). *Evocative objects: Things we think*. Cambridge, MA: The MIT Press.
Uit Vrije Wil [Out of Free Will]. (2017). *Burgerinitiatief voltooid leven* [Citizen initiative on life completed]. [Official website]. Accessed on March 29, 2022 from http://uitvrijewil.nu/index.php?id=1000.
Ureta, J., Brito, C. I., Dy, J. B., Santos, K. A., Villaluna, W., & Ong, E. (2020). At home with Alexa: A tale of two conversational agents. In P. Sojka, I. Kopecek, K.

Pala, & A. Horak (Eds.), *International conference on text, speech, and dialogue proceedings* (pp. 495–503). Cham: Springer.

Valkenberg, S. (2014, April). Kijkje in de zaadcel: Wat doen we ermee? [A glance at the sperm cell: What do we do with it?] *Trouw*. Accessed on March 29, 2022 from https://www.trouw.nl/nieuws/kijkje-in-de-zaadcel-wat-doen-we-ermee~b96ad47f/.

Vallor, S. (2016). *Technology and the virtues: A philosophical guide to a future worth wanting*. Oxford: Oxford University Press.

Value, aesthetic. (2005). In T. Hondreich (Ed.), *The Oxford companion to philosophy* (2nd ed., p. 941). Oxford: Oxford University Press.

Value. (2005). In T. Hondreich (Ed.), *The Oxford companion to philosophy* (2nd ed., p. 941). Oxford: Oxford University Press.

Van Dam, K. H., Nikolic, I., & Lukszo, Z. (Eds.). (2012). *Agent-based modelling of socio-technical systems* (Vol. 9). Springer Science & Business Media.

Van de Poel, I. (2009a). The introduction of nanotechnology as a societal experiment. In S. Arnaldi, A. Lorenzet, & F. Russo (Eds.), *Technoscience in progress. Managing the uncertainty of nanotechnology* (pp. 129–143). Amsterdam: IOS Press.

Van de Poel, I. (2009b). Values in engineering design. In A. W. M. Meijers (Ed.), *Philosophy of technology and engineering sciences* (pp. 973–1006). Amsterdam, North Holland: Elsevier.

Van de Poel, I. (2011). Nuclear energy as a social experiment. *Ethics, Policy & Environment, 14*(3), 285–290.

Van de Poel, I. (2013). Why new technologies should be conceived as social experiments. *Ethics, Policy & Environment, 16*(3), 352–355.

Van de Poel, I. (2015). Design for values. In P. Kawalec & R. P. Wierzchoslawski (Eds.), *Social Responsibility and Science in Innovation Economy* (pp. 115–165). Lublin: Learned Society of KUL & John Paul II Catholic University of Lublin.

Van de Poel, I. (2016). An ethical framework for evaluating experimental technology. *Science and Engineering Ethics, 22*(3), 667–686.

Van de Poel, I. (2021). Design for value change. *Ethics and Information Technology, 23*(1), 27–31.

van de Poel, I., & Kudina, O. (2022). Understanding technology-induced value change: A pragmatist proposal. *Philosophy & Technology, 35*(2), 40.

Van den Eede, Y. (2015). Tracing the tracker: A postphenomenological inquiry into self-tracking technologies. In R. Rosenberger & P.-P. Verbeek (Eds.), *Postphenomenological investigations: Essays on human–technology relations* (pp. 143–158). Lanham: Lexington Books.

Van den Hoven, J. (2005). Design for values and values for design. *Information Age, 4*, 4–7.

Van den Hoven, J., & Manders-Huits, N. (2009). Value-sensitive design. In J. K. Berg Olsen, S. A. Pedersen, & V. F. Hendricks (Eds.), *A companion to the philosophy of technology* (pp. 477–780). Wiley Online Library. https://doi.org/10.1002/9781444310795.ch86.

Van den Hoven, J., & Vermaas, P. E. (2007). Nano-technology and privacy: On continuous surveillance outside the panopticon. *Journal of Medicine and Philosophy, 32*(3), 283–297.

Van Hoof, W., Pennings, G., & De Sutter, P. (2015). Cross-border reproductive care for law evasion. *European Journal of Obstetrics & Gynecology and Reproductive Biology, 2002*, 101–105.

Van Wynsberghe, A. (2012). *Designing robots with care: Creating an ethical framework for the future design and implementation of care robots* (PhD thesis). Enschede: University of Twente.

Van Wynsberghe, A., & Robbins, S. (2014). Ethicist as designer: A pragmatic approach to ethics in the lab. *Science and Engineering Ethics, 20*(4), 947–961.

Veen, M., Gremmen, B., te Molder, H., & van Woerkum, C. (2011). Emergent technologies against the background of everyday life: Discursive psychology as a technology assessment tool. *Public Understanding of Science, 20*(6), 810–825.

Verbeek, P.-P. (2003). Material hermeneutics. *Techné: Research in Philosophy and Technology, 6*(3), 181–184.

Verbeek, P.-P. (2005). *What things do: Philosophical reflections on technology, agency, and design*. Penn State University Press.

Verbeek, P.-P. (2006). Materializing morality: Design ethics and technological mediation. *Science, Technology & Human Values, 31*(3), 361–380.

Verbeek, P. P. (2008). Obstetric ultrasound and the technological mediation of morality: A postphenomenological analysis. *Human Studies, 31*(1), 11–26.

Verbeek, P.-P. (2010). Accompanying technology: Philosophy of technology after the ethical turn. *Techné: Research in Philosophy and Technology, 14*(1), 49–54.

Verbeek, P.-P. (2011). *Moralizing technology: Understanding and designing the morality of things*. University of Chicago Press.

Verbeek, P.-P. (2014). Some misunderstandings about the moral significance of technology. In P. Kroes & P.-P. Verbeek (Eds.), *The moral status of technical artefacts* (pp. 75–88). Dordrecht: Springer.

Verbeek, P.-P. (2015). Toward a theory of technological mediation: A program for postphenomenological research. In J. K. Berg O. Friis & R. P. Crease (Eds.), *Technoscience and postphenomenology: The Manhattan papers* (pp. 189–205). Lanham: Lexington Books.

Verbeek, P.-P., Brey, P., van Est, R., van Gemert, L., Heldeweg, M., & L. Moerel. (2020, September 10). Ethische analyse van de COVID-19 notificatie-app ter aanvulling op bron en contactonderzoek GGD. Deel 2: analyse door burgers en professionals. [Trans.: Ethical analysis of the COVID-19 notification app to supplement epidemiological source and contact research. Part 2: Analysis by citizens and professionals]. *Rijksoverheid.* Accessed on October 19, 2021 from https://ecp.nl/wp-content/uploads/2020/11/rapport-begeleidingsethiek-coronamelder.pdf.

Volkskrant. (1996, February 27). Meerderheid bevolking wil verbod geslachtskeuze baby in kliniek [The majority of population wants a ban on the sex selection of babies in clinic]. *Volkskrant* [web archive]. Accessed on March 29, 2022 from https://www.volkskrant.nl/nieuws-achtergrond/meerderheid-bevolking-wil-verbod-geslachtskeuze-baby-in-kliniek~be2fac51/.

Wallach, W., & Allen, C. (2009). *Moral machines: Teaching robots right from wrong*. New York: Oxford University Press.

Walzer, M. (1994). *Thick and thin: Moral argument at home and abroad.* Notre Dame: University of Notre.

Wayback Machine. (2015). Glass. What it does. *The Internet Archive.* Accessed on April 5, 2022 from https://bit.ly/2QqwcMv.

Wellner, G. (2015). *A postphenomenological inquiry of cell phones: Genealogies, meanings, and becoming.* Lanham: Lexington Books.

West, M., Kraut, R., & Chew, H. E. (2019). *I'd blush if I could: Closing gender divides in digital skills through education* [Report]. UNESCO & EQUALS Skills Coalition.

Wiederhold, B. K. (2018). "Alexa, are you my Mom?" The role of artificial intelligence in child development. *Cyberpsychology, Behavior, and Social Networking, 21*(8), 471–472.

Wikler, D., & Wikler, N. J. (1991). Turkey-baster babies: The demedicalization of artificial insemination. *The Milbank Quarterly, 69*(1), 5–40.

Williams, E. (2020, August 3). COVID tracing apps: What Europe has done right, and wrong. *Hackaday.* Accessed on December 24, 2021 from https://hackaday.com/2020/08/03/covid-tracing-apps-what-europe-has-done-right-and-wrong/.

World Health Organization. (2011). *Preventing gender-biased sex selection.* Geneva: WHO Press.

Wu, Y., Rough, D., Bleakley, A., Edwards, J., Cooney, O., Doyle, P. R., ... Cowan, B. R. (2020). See what I'm saying? Comparing Intelligent Personal Assistant use for native and non-native language speakers. In *22nd International Conference on Human-Computer Interaction with Mobile Devices and Services* (Article 34, pp. 1–9). New York: Association for Computing Machinery.

Yetim, F. (2011). Bringing discourse ethics to value sensitive design: Pathways toward a deliberative future. *AIS Transactions on Human-Computer Interaction, 3*(2), 133–155.

Index

Aagaard, J., 8, 9
AI-powered voice assistants, 13, 117–18
appropriation: definition, 5, 49–50; lens of domestication study, 50–55
Aristotle, 22, 26
assisted reproduction technology, 1

"bodies," 109
Body One, 109–11
Body Two, 109–11
Boenink, M., 100, 131
Borgmann, A., 107, 108

Capurro, R., 107, 108
care-centered VSD approach, 126
Carter, S., 50, 51
classical phenomenology, 30n2
cold ethics, 35
Collingridge dilemma, 128–30, 132, 133, 135, 138
communication privacy, 62–63
Constructive Technology Assessment, 128
contact-tracing apps, 17
Conversation Analysis and Discursive Psychology (CA&DP), 73, 100, 131; exploring of, 73–75; *vs.* IPA, 78–81
Corona apps. *See* COVID-19 tracking apps
co-shaping relationship, 38
COVID-19 tracking apps, 16–18, 137

decision-making, 15, 18, 20, 28; moral infrastructure, 32
deliberation, 47n2
destabilization, of moral routines, 34
Dewey, J., 15, 19, 20, 22–28, 30n1, 31, 34–36, 38, 39, 45, 46, 46n1, 47nn2, 3
digital voice assistants (DVAs), 103–4, 106, 110, 111, 113–17; preliminary examination of, 104; privacy either-or model of, 118
discursive methods, 75
domestication: lens of, 50–52; technological mediation, 52–55
double hermeneutics, 76, 77
Dreyfus, H. L., 107, 108
DVAs. *See* digital voice assistants (DVAs)

empirical philosophy, 9, 20, 58, 71, 82
empirical postphenomenology: current status of, 7–9; to study moral hermeneutics, 9–10
Encyclopedia of Philosophy, 20
ethical decision-making, 60
ethics, 35, 45; as accompaniment, 28; Collingridge dilemma in, 133; role of, 28
experimental empiricism, 22, 23

Flowers, P., 72, 77, 82
Friedman, B., 122, 123, 126

Gadamer, H.-G., 25, 72, 75, 76, 79, 104, 108, 113–15, 118, 118n3, 137; circular account, 105–7, *106*
genetic disorders, 85
Glass Explorers, 135
"Glass-free zone," 57
"Glassholes," 59
Google Cloud, 59
Google Glass, 9–10, 38, 55–58, 71, 84, 126, 130, 134, 139; back in control of your technology, 58–61; preliminary appropriation study of, 68; preliminary investigation of, 69; reasoning with privacy, 61–66; reflecting on preliminary study, 67–68
Green, J., 50, 51
Grunwald, A., 49, 53

Haen, D., 74, 131
hermeneutic circle account, 105–7, *106*, 115, 137
hermeneutic lemniscate, 104, 115, 118, 137; mediated meaning as, *112*, 112–15
hermeneutics, 72, 75, 79, 105–6, 114, 115, 118; developments in, 108; and Gadamer's circular account, 105–7, *106*
HoloLens, 58
human rationality, 21–22
human-technology intertwinement, 4
human-technology relations, 6, 8
human-technology-world relations, 4, 7–9, 25, 28, 32, 35, 38, 45, 46, 112, 118n5, 139; types of, 109
human-world relations, 107–12, *110*
Husserl, E., 19, 29n1, 30n1
Husserlian phenomenology, 30n1

idiographic research, 101n1
Ihde, D., 19, 20, 29n1, 30n2, 104, 107, 118n5; human-world relations, 107–12, *110*

information and communication technology (ICT) community, 123
Internet-based communication, 2
Interpretative Phenomenological Analysis (IPA), 73, 100, 126, 127, 133, 138; *vs.* CA&DP, 78–81; exploring of, 75–78; into method for studying moral hermeneutics, 81–84
interrelational ontology, 30n2
IPA-based mediation analysis, 138

Kant, I., 37, 47n3
Kudina, O., 62, 67, 100

Larkin, M., 72, 77, 82
Le Dantec, C. A., 123, 124
lemniscatic principle of interpretation, 31–32, 114, 139; distinctions between technomoral change and, 41; scope of, 35
liberalism, 91–97
Locke, J., 19–20

macro-level of technomoral change, 39–40
mass media, global penetration of, 44
material hermeneutics, 109–10, *110*, 114, 118n5
Mayer-Schönberger, V., 65
meaning-making process, 72, 75, 80
mediation approach, 4, 6, 24, 26, 27, 32, 38, 40
medical imaging technologies, 3
meso-level of technomoral change, 40
microfluidic chip-based technology, 86
micro-level of technomoral change, 40
mixed-reality glasses, 57–58
Mol, A., 46, 58
moral actions, 116
moral disagreement, 34, 47n3
moral economy, 51
moral hermeneutics, 4, 5, 45, 49, 100, 121, 136, 138, 140; comprehensive study of, 68–69; designing technologies with,

122–27; dynamics of, 80; empirical inquiry into, 5–7; empirical postphenomenology, 7–10; IPA into method for studying, 81–84; perspective of, 125; philosophical investigation of, 71–73; research direction for, 10–13; sex selection and, 84–99; technological innovation and, 129; and technologies, 115–17; technologies in, 39–45; and technology governance, 127–36

morality, 45; dynamic nature of, 2; levels of, 37; as "moral routines" and "moral habits," 34; role of technologies in, 1–5; technological mediation, levels and stages of, 35–39

morality-in-the-making, 49–50, 54, 68, 99

moral mediation, 3–5; mediation part in, 31–32; moral part of, 15–16

moral mediation account, 37

"moral mediators," 2

moral perceptions, 108, 115–16

moral philosophy, 22; of technology, 29

moral sense-making, 4, 10–13, 69, 73, 99–101, 103, 116–18, 121, 125, 126, 133, 136, 137, 139

Multi-Agent Systems Modelling/Choice Behavior Modelling, 140

natural language processing algorithms, 103
non-users, 28, 29

Oxford Dictionary of Philosophy, 20
Oxford Dictionary of Sociology, 21

philosophical inquiry, 27, 28
philosophy, role of, 28
policy-makers, 29
postphenomenology, 1; and Ihde's human-world relations, 107–12, *110*; political dimension of, 7
potentiality, 36
practice-based approach, 19
pragmatism, 19, 20, 30n2

pre-implantation genetic diagnosis (PGD), 85
prejudice, 106
privacy, 56, 137; and attention, 62; as civil inattention, 66; of communication, 62–63; of experience and memories, 64–65; as limited access to self, 63–64; preliminary investigation of, 69; in public space, 65–66; reasoning with, 61–62; values of, 57–68, 126
Project Alias, 116
projective appropriation, 52
proof of principle, need for, 55–57
psychology, 82

reflective deliberation, 47n2
reflective inquiry, scholarship on, 26
relationality, 24–25
Rip, A., 34, 35, 38
Rosenberger, R., 7–9

Science and Technology Studies (STS), 19, 50
self-tracking technologies, 8, 32
sense-making activity, 54, 73, 77–81, 104, 113
sense-making process, 73, 75; hermeneutics and Gadamer's circular account, 105–7, *106*; postphenomenology and Ihde's human-world relations, 107–12, *110*
sex selection technology (SST), 10, 84–85, 135; chip-based form of, 86; ethical debate, 87–89; IPA findings, 91–99; setup of IPA study, 89–91; technological background, 85–87
smart speakers, 103
Smith, J. A., 72, 76, 77, 82
Social Construction Theory, 50
sociotechnical experimentation approach, 128, 132–33, 139
Spectacles, 58
speech recognition systems, 111
sperm sorting, 85

SST. *See* sex selection technology (SST)
SST+, 84–87; desirability of, 95; direct-to-consumer technology, 88; in Dutch society, 98; moral hermeneutics potential of, 89; on parental values and parent-child relations, 92; in relation to values of parenthood, 91
Swierstra, T., 31, 34–35, 38–39, 41, 44, 54, 128, 131

technological appropriation, 103, 104; testing assumptions of, 57–68
technological mediation approach, 19, 33, 34, 39–45, 52–55, 79, 80, 115, 129, 133–36; different levels and stages of morality, 35–39; moral hermeneutics perspective in, 125
technology: designing with moral hermeneutics, 122–27; ethical dimension of, 2, 15; ethical implications of, 2–3; interrelation of, 121; in moral hermeneutics, 39–45; moral hermeneutics and, 115–17; moral philosophy of, 29; moral significance of, 32; perception-action mediation of, 107–8; rise of, 1; role in co-shaping human relations with world, 32; role in mediation process, 6; role in morality, 1–5
technology governance: moral hermeneutics and, 127–30; sociotechnical experiments, 132–33; technological mediation, 133–36; technomoral change, 130–32
technomoral change approach, 2, 39–45, 130–32, 141n3
technomoral scenarios approach, 129–32
telecare technologies, 3
theory of valuation, 23
Thorogood, N., 50, 51
tripartite methodology, 124

uncertainty, 26
urban technology, 7

value adaptation, 43
value change approach, 39–45
value conceptualization, 41, 42
value dynamism, 3, 4, 6–7, 9, 10, 16, 25, 33, 34, 38, 43, 133
value formation, 22
values: comprehensive list of, 140n1; definitions of, 20–21; developmental and relational account of, 136; dictionaries and formal theories of, 20; dimensions of, 41; ends-in-view, 23–24; examining through formal theories, 20–24; interrelation of, 121; postphenomenological on, 18–20; pragmatist definition of, 24; preliminary practice-based examination of, 16–18; of privacy, 57–68, 126; relational and dynamic view on, 33–35; relational conception of, 25; sensitivity to, 125–26; solidifying perspectives through pragmatism, 24–29
Value Sensitive Design (VSD), 121–27, 139, 140n1
values-in-practices, 20
"value suitabilities," 125
Van de Poel, I., 22, 31, 39, 44, 62, 67, 121, 128, 132, 133
Van Wynsberghe, A., 124, 126
Verbeek, P.-P., 3, 5, 8, 9, 15, 18, 26–28, 31, 34, 41, 49, 54, 73, 81, 110, 111, 125, 135–37; co-shaping idea, 115; *What things do?*, 109
voice assistants process, 103–4
voice-based interface, 106–8, 110
VSD. *See* Value Sensitive Design (VSD)

Walzer, M., 31, 35, 37, 38
Wellner, G., 7, 32
What things do? (Verbeek), 109

YouTube, 57, 58, 60, 61, 67, 70n2, 71, 72, 135; mediation analysis, 134

About the Author

Olya Kudina is an assistant professor in ethics/philosophy of technology, exploring the dynamic interaction between values and technologies. She combines the phenomenological and pragmatist focus with cultural sensitivity to study morality as an evolving system. Her expertise in empirical philosophy helps her connect ethics and design in fostering responsible human-AI collaborations, with a recent focus on AI in (mental) healthcare. Olya holds a PhD degree in Philosophy of Technology from the University of Twente. Her previous work outside academia adds to her skill-set in the areas of diplomacy, (inter)governmental work, data protection, and privacy.

Milton Keynes UK
Ingram Content Group UK Ltd.
UKHW031922041124
450718UK00006B/58